# Advanced Computer Architectures

To
Kalpana
Sruti and Sweta

# Preface

Advances in hardware and software technologies have enabled us to build very powerful computer systems. As the capability of these systems grew, the demand for more powerful systems to enable a variety of applications also has grown. It is ideal to build a single processor system that can address the needs of any application, since that architecture eliminates most of the overhead required to build systems with multiple processors. Since, it is not always practical to do so, architectures with a multiplicity of processors interconnected in multiple ways have been utilized. This book addresses the fundamental aspects of such systems. It is my belief that a good understanding of the capabilities and limitations of underlying hardware structures makes an application developer more efficient. Thus, this book follows the method of introducing hardware structures for pipelined and parallel processors first, followed by the concepts of software structures and algorithms.

This book is designed for use in a second course in computer architecture typically offered by computer science, computer engineering and electrical engineering departments. Such a course would typically be at an advanced senior or beginning graduate level. This book also satisfies the 'Advanced topics in Architecture' portion of the IEEE CS Curricula 2001. A course in computer architecture and organization is a prerequisite for this book. I have included a review of the prerequisite material in the Introduction and Chapter 1. Chapter 2 introduces the most popular processing paradigms and the corresponding terminology. Performance evaluation and cost factor considerations are outlined in this chapter. Chapter 3 introduces pipeline design and Chapter 4 extends the pipeline concepts to describe vector processors, a heavily pipelined architecture type. Chapter 5 describes data parallel architectures more commonly known as array processors. Multiprocessor systems are described in Chapter 6. Chapter 7 describes some of the current directions in computer architecture to include dataflow, grid, Biology inspired and optical computing. Dataflow architectures have been of experimental interest over the last few years, although no major dataflow architecture is commercially available now. With the advent of internet, cluster computing and more recently grid computing has become practical and seems to be the current most popular paradigm. An array of architectures inspired by Biology (Neural networks, DNA computing and artificial immune systems) has evolved over the years. While optical components are being extensively used in data communication, optical computing is still in its nascent state.

This book is a result of my teaching a sequence of courses in computer architecture to computer science and computer engineering students over the last thirty years. The sequence started with a course in logic design and ended with a graduate course in distributed processing. I have used the material in this book in the second course on architecture. The course used Chapter 1 as a review and covered Chapters 2 through 6, ending with selected topics from Chapter 7. In addition to covering systems given as examples in the book, I have assigned studying other contemporary systems as case studies in the course.

I have tried to include the details of the most appropriate commercially available systems as examples in each chapter of the book. The field of parallel processing has seen a continuous experimentation in building systems. Most of these systems have either remained laboratory curiosities or produced in small quantities and have been made obsolete. As such, I have stayed away from complete descriptions of experimental architectures, although I have utilized the pertinent concepts used by these systems to reinforce the material presented. One problem with using commercially available systems as examples is that, they become obsolete by the time the book is published. The only solution for this is to refer the reader to the literature from the manufacturers of these systems for the latest details.

I have made every effort to credit the source for the copyrighted material used in this book. My apologies to any copyright holders whose rights I have unwittingly infringed. My thanks to all the manufacturers of machines I have used as examples in the book for their permission to use the material.

Many individuals have influenced the development of this book. I thank all the students in my architecture courses over the last few years for their thoughts, comments, suggestions, and tolerance during the evolution of this book. Thanks to Sakshi Sethi and Sai Kolli for their support in manuscript preparation and Dr. Sruti Shiva for her suggestions on Biology inspired architectures. It is a pleasure to acknowledge the suggestions and encouragement from the readers of my earlier books. They made this book better and working on it enjoyable. Thanks are also due to the anonymous reviewers for their suggestions to make this book what it is today.

It is my pleasure to acknowledge the encouragement from Russell Dekker of Marcel-Dekker in starting this project. My thanks to Mohan and his staff at Newgen Imaging, and Suzanne Lassandro and Jessica Vakili at Taylor and Francis for their superb support in the production of the book.

My family put up with me one more time and allowed me to spend long hours hiding in my study during the writing of this book. I thank my wife Kalpana and daughters Sruti and Sweta for their love, understanding and undying support.

Sajjan G. Shiva

# The Author

**Sajjan G. Shiva, Ph.D.,** is Professor and Chairman of the Computer Science Department at the University of Memphis. Dr. Shiva received his Ph.D. in Electrical Engineering from Auburn University in 1975. Previously he served on the faculty of the computer science departments at the University of Alabama, Huntsville and Alabama A&M University. Dr. Shiva has taught and conducted research in the areas of software engineering, computer architecture, distributed and parallel processing, and artificial intelligence. He has also served as a consultant to industry and government since 1975, and is a Fellow of IEEE. Dr. Shiva's books on computer architecture have been adopted by more than 87 universities in the U.S., Canada, Egypt, China, India, Australia, and other countries.

# Contents

**Introduction**   1
   I.1   Computing Paradigms   2
   I.2   The Need for Advanced Architectures   6
       I.2.1   Hardware   7
       I.2.2   Software   9
       I.2.3   Systems   10
   I.3   Book Overview   13
   References   14

**1 Uniprocessor Architecture Overview**   15
   1.1   Uniprocessor Model   15
   1.2   Enhancements to the Uniprocessor Model   18
       1.2.1   Arithmetic Logic Unit   19
       1.2.2   Memory   20
       1.2.3   Control Unit   23
       1.2.4   I/O Subsystem   26
       1.2.5   Interconnection Structures   27
       1.2.6   System Considerations   28
   1.3   Example Systems   29
       1.3.1   Intel Corporation's Itanium   29
       1.3.2   MIPS Computer System's R10000   38
   1.4   Summary   45
   Problems   45
   References   47

**2 Models and Terminology**   49
   2.1   Effect of Application on the Architecture   49
   2.2   Application Characteristics   50
   2.3   Processing Paradigms   52
   2.4   Flynn's Taxonomy   55
       2.4.1   Single Instruction Stream, Multiple Data Stream   59
       2.4.2   Multiple Instruction Stream, Multiple Data Stream   60
   2.5   Computer Networks   63
   2.6   Performance Evaluation   64
       2.6.1   Benchmarks   65
   2.7   Cost Factor   68
   2.8   Summary   70
   Problems   70
   References   71

## 3 Pipelining — 73

- 3.1 Pipeline Model — 73
  - 3.1.1 Pipeline Types — 77
- 3.2 Pipeline Control and Performance — 85
  - 3.2.1 Collision Vectors — 88
  - 3.2.2 Control — 90
  - 3.2.3 Performance — 91
  - 3.2.4 Multifunction Pipelines — 94
- 3.3 Other Pipeline Problems — 96
  - 3.3.1 Data Interlocks — 96
  - 3.3.2 Conditional Branches — 102
  - 3.3.3 Multiple Instruction Buffers — 106
  - 3.3.4 Interrupts — 107
- 3.4 Dynamic Pipelines — 107
  - 3.4.1 Instruction Deferral — 108
  - 3.4.2 Performance Evaluation — 113
- 3.5 Example Systems — 114
  - 3.5.1 Control Data Corporation STAR-100 — 114
  - 3.5.2 Control Data Corporation 6600 — 115
  - 3.5.3 Sun Microsystem's Niagara Microprocessor — 117
- 3.6 Summary — 119
- Problems — 119
- References — 122

## 4 Vector Processors — 123

- 4.1 Vector Processor Models — 124
- 4.2 Memory Design Considerations — 128
- 4.3 Architecture of the Cray Series — 133
  - 4.3.1 Memory — 133
  - 4.3.2 Processor Interconnection — 134
  - 4.3.3 Central Processor — 134
  - 4.3.4 I/O System — 143
  - 4.3.5 Other Systems in the Series — 144
- 4.4 Performance Evaluation — 144
- 4.5 Programming Vector Processors — 147
- 4.6 Example Systems — 152
  - 4.6.1 Hitachi Super Technical Server — 152
  - 4.6.2 NEC SX Series — 157
  - 4.6.3 Cray X1 — 160
- 4.7 Summary — 162
- Problems — 163
- References — 165

## 5 Array Processors — 167
- 5.1 SIMD Organization — 167
  - 5.1.1 Memory — 169
  - 5.1.2 Control Processor — 169
  - 5.1.3 Arithmetic/Logic Processors — 170
  - 5.1.4 Interconnection Network — 170
  - 5.1.5 Registers, Instruction Set, Performance Considerations — 172
- 5.2 Data Storage Techniques and Memory Organization — 175
- 5.3 Interconnection Networks — 182
  - 5.3.1 Terminology and Performance Measures — 182
  - 5.3.2 Routing Protocols — 185
  - 5.3.3 Static Topologies — 187
  - 5.3.4 Dynamic Topologies — 192
- 5.4 Performance Evaluation and Scalability — 199
- 5.5 Programming SIMDs — 204
- 5.6 Example Systems — 204
  - 5.6.1 ILLIAC-IV — 205
  - 5.6.2 Thinking Machine Corporation's CM-2 — 207
- 5.7 Summary — 217
- Problems — 218
- References — 220

## 6 Multiprocessor Systems — 221
- 6.1 MIMD Organization — 224
  - 6.1.1 Shared-Memory Architecture — 224
  - 6.1.2 Message-Passing Architecture — 226
  - 6.1.3 Other Models — 226
- 6.2 Memory Organization — 227
  - 6.2.1 Cache Coherence — 229
- 6.3 Interconnection Networks — 233
  - 6.3.1 Bus Network — 235
  - 6.3.2 Loop or Ring — 237
  - 6.3.3 Mesh Network — 237
  - 6.3.4 Hypercube Network — 238
  - 6.3.5 Crossbar Network — 238
  - 6.3.6 Multistage Networks — 238
- 6.4 Operating System Considerations — 242
  - 6.4.1 Synchronization Mechanisms — 246
  - 6.4.2 Heavy- and Light-Weight Processes — 252
  - 6.4.3 Scheduling — 254
- 6.5 Programming — 255
  - 6.5.1 Fork and Join — 257
  - 6.5.2 Parbegin/Parend — 259
  - 6.5.3 DOALL — 259
  - 6.5.4 Shared Variables — 260
  - 6.5.5 Send/Receive — 260

|     |       |                                           |     |
|-----|-------|-------------------------------------------|-----|
| 6.6 |       | Performance Evaluation and Scalability    | 260 |
|     | 6.6.1 | Scalability                               | 262 |
|     | 6.6.2 | Performance Models                        | 263 |
| 6.7 |       | Example Systems                           | 268 |
|     | 6.7.1 | Thinking Machine Corporation's CM-5       | 268 |
|     | 6.7.2 | Cray Research Corporation's T3D           | 270 |
|     | 6.7.3 | IBM System X                              | 274 |
| 6.8 |       | Summary                                   | 276 |
|     |       | Problems                                  | 277 |
|     |       | References                                | 280 |

# 7 Current Directions — 283

| 7.1 |       | Dataflow Architectures              | 284 |
|-----|-------|-------------------------------------|-----|
|     | 7.1.1 | Dataflow Models                     | 285 |
|     | 7.1.2 | Dataflow Graphs                     | 290 |
|     | 7.1.3 | Dataflow Languages                  | 292 |
|     | 7.1.4 | Example Dataflow Systems            | 294 |
| 7.2 |       | GRID Computing                      | 299 |
|     | 7.2.1 | Open Grid Services Architecture     | 302 |
| 7.3 |       | Biology Inspired Computing          | 305 |
|     | 7.3.1 | Neural Networks                     | 305 |
|     | 7.3.2 | DNA Computing                       | 317 |
|     | 7.3.3 | Artificial Immune Systems           | 320 |
| 7.4 |       | Optical Computing                   | 322 |
|     |       | References                          | 327 |

Index — 331

# Introduction

This book is about the architecture of *high-performance* computer systems. Architecture is the "art or science of building; a method or style of building" according to Webster's. Computer architecture comprises both the art and science of designing new and faster computer systems to satisfy the ever-increasing demand for more powerful systems. A computer architect specifies the modules that form the computer system at a functional level of detail and also specifies the interfaces between these modules. Traditionally, these modules were thought to be hardware elements consisting of processors, memories, and input/output (I/O) devices. With the advances in hardware and software technologies it is now possible to formulate efficient structures for each module using hardware, software, firmware, or a combination of the three. As such, at the architecture level, consideration of the detailed functional characteristics of each module is more important than the implementation details of the module. The exact mix of hardware, software, and firmware used to implement the module depends on the performance requirements, cost, and the availability of the hardware, software, and firmware components.

*Performance* and *cost* are the two major parameters for evaluation of architectures. The aim is to maximize the performance while minimizing the cost (i.e., maximize the *performance-to-cost ratio*). Performance is usually measured as the maximum number of operations performed per second, may they be arithmetic operations, logical inferences, or I/O operations. In addition to these, other measures such as maximum program and data size, ease of programming, power consumption, weight, volume, reliability, and availability have also been used to evaluate architectures.

Progress in hardware and software technologies provides new choices each year. The architect has to not only base his or her decisions on the choices available today, but also keep in mind the expected changes in technology during the life of the system. An architectural feature that provides the optimum performance-to-cost ratio today may not be the best feature for tomorrow's technology.

Most of the performance enhancements achieved in modern day systems have been due to advances in hardware technology. In fact, the four *generations* of computer systems are identified with the improvement in hardware technology, viz. vacuum tubes, transistors, medium-scale integrated (MSI) circuits, and large-scale integrated (LSI) circuits. It is also interesting to note that the first three *waves* of computing are also based on hardware technology.

The first wave in the 1960s was dominated by mainframes. The second wave in the 1970s belonged to the minicomputer systems. The third wave in the 1980s was that of microcomputers. Since these generations and waves were brought about primarily based on the improvements in hardware technology, they are considered *evolutionary* approaches to building high-performance machines.

While early generations of computer hardware were more-or-less accompanied by corresponding changes in computer software, since the mid-1970s, hardware and software have followed divergent paths. While the introduction of workstations and personal computers have changed the way hardware and software are made available to end users, programming methodologies and the types of software available to end users have followed a path independent of the underlying hardware.

It is often said that the limits of semiconductor technology have already been reached and no significant performance improvements are possible by technology improvements alone. History has always proven this statement to be false. But, we always make the most out of the existing technology. When the technology improvements alone cannot yield the desired performance, alternative-processing architectures that produce performance improvements based on the existing technology without the need to "wait for the next generation" should be deployed. The alternative architectures utilize the parallel and overlapped (pipelined) processing possibilities.

Parallel processing is considered as the *fourth wave* of computing. Parallel-processing architectures utilize a multiplicity of processors and provide for building computer systems with orders-of-magnitude performance increases. In order to utilize parallel-processing architectures efficiently, an extensive redesign of algorithms and data structures (as utilized in the nonparallel implementation of the application) is needed. As such, these are considered *revolutionary* approaches for high-performance computer system architecture.

This book covers both evolutionary and revolutionary approaches. Section I.1 of this chapter provides a definition of the three common computing *paradigms* (models). Section I.2 highlights the need for advanced architectures, and Section I.3 provides the overview of the book.

## I.1 Computing Paradigms

Computing structures can be broadly represented by three paradigms: *serial, pipelined, and parallel*. These paradigms are illustrated by the following example:

> **Example I.1**
> Figure I.1 shows the three models with reference to the task of building an automobile. This task can be very broadly partitioned into

# Introduction

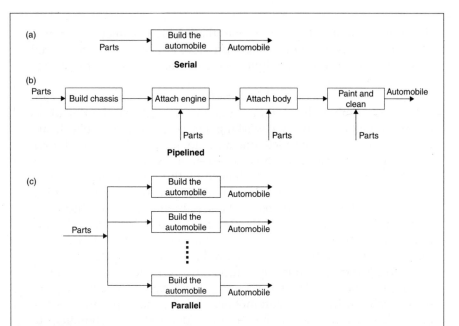

**FIGURE I.1**
Three paradigms: (a) serial, (b) pipelined, and (c) parallel

the following subtasks: assembling the chassis, attaching the engine, attaching the body, and painting and cleaning.

Figure I.1(a) shows the *serial* paradigm. Here all the four subtasks are performed at one station. That is, there is a single server and the four subtasks are done in sequence for each automobile. Only one automobile is worked on at a time. If each of these subtasks takes $T$ minutes to complete, $4T$ minutes are needed to assemble each automobile and hence, $4NT$ minutes are needed to assemble $N$ automobiles.

Figure I.1(b) shows the *pipelined* processing paradigm, which resembles the standard automobile assembly line. Here, the assembly is done in four stages. Each stage is specialized to perform one subtask. Once an automobile leaves stage 1 and moves to stage 2, the next automobile will be worked on by stage 1. The other three stages also work in this *overlapped* manner. That is, it takes $4T$ minutes to fill the assembly line with the work. The first automobile is complete at the end of $4T$ minutes and then onward one automobile is completed every $T$ minutes. It still takes $4T$ minutes to assemble an automobile. But, to assemble $N$ automobiles, it takes $4T + (N-1)T$ minutes. The *speedup* achieved by the pipeline is, thus:

$$\frac{4NT}{4T + (N-1)T} \tag{I.1}$$

For large N, the speedup is four times that of serial architecture. In general, an ideal pipeline with P stages provides a speed up of P compared to a serial architecture. This ignores the overhead required to manage the stages of the pipeline.

Figure I.1(c) shows the *parallel* paradigm. Here, P stations, each capable of performing all the four subtasks are used. All the P stations operate simultaneously, each working on a separate automobile. It still takes 4T minutes to complete one automobile, but P automobiles are produced in 4T minutes. Thus, to assemble N automobiles it takes 4NT/P minutes. The speed up achieved by this architecture is P times that of the serial architecture. Note that the hardware complexity has increased by P-fold compared to the serial paradigm, to achieve the P-fold speed up. Also, each of the P stages could have been implemented, as a pipeline as in Figure I.1(b), thus enhancing the throughput further, although at the cost of increased hardware complexity.

The four subtasks in this application must be performed in sequence to complete the task. There is no way to perform all the subtasks simultaneously on the same automobile. That is, the application is not parallelizable and hence we had to use P identical production lines to enhance the throughput.

**Example I.2**

As another example, consider the computation of the column-sum of an $(N \times N)$ matrix. In the serial (single-processor) implementation of this task, the processor takes N time units to compute the sum of each column, and hence requires $N^2$ time units to complete the task. In the parallel implementation of this task, N processors are used. Each processor accumulates elements of a column, in N time units. Since, all the processors operate simultaneously, the total time for the completion of the task is N. Thus, the throughput is enhanced N-fold.

The pipelined- and parallel-processing architectures in general, provide a higher throughput compared to serial architecture. These architectures utilize multiple processors and are more complex compared to serial architectures.

The key attribute of the application is that it should be partitionable into subtasks that can be executed simultaneously, to make the parallel processing possible. When such a partitioning is not possible (as in the automobile assembly example above), the pipeline mode of computation is adopted. The matrix computation example above, is the case of the so-called *trivial parallelism* in which the job could be partitioned into independent subtasks and the subtasks do not exchange data (i.e., communicate with each other) during the computation. In practice, such clean partitioning is not always possible and there will be a considerable communication and task coordination overhead associated with the parallel and pipelined structures which reduce their throughput. This overhead is ignored in the above illustrations. The overall processing speed thus depends on two factors: the *computing* speed of the processors and the *communication* speed of their interconnection structure. The communication overhead is a function of the *bandwidth* (i.e., the number of units of data that can be transmitted per second) and the *latency* (how long it takes to transmit data from a source to a destination) of the interconnection structure.

Obviously, it is better to implement the application on a single-processor architecture, if it provides the required performance in a cost-effective manner, since it eliminates the communication overhead. If not, multiple-processor architectures operating in pipelined and parallel modes should be utilized and the communication overhead should be minimized. The supercomputers of the 1990s provided for this implementation mode for applications that cannot be easily partitioned into a large number of parallel subtasks. They offered very high speeds utilizing relatively a small number of functional units (4–16) and extensive pipelining. The supercomputers of today have large number of processors.

The performance/cost characteristics of serial computers over the last few decades have shown an interesting trend. The characteristic remains linear at the lower performance levels. As the performance level increases, the cost becomes exorbitantly high for small increments in performance. With the advances in very large-scale integrated circuit (VLSI) technology, it is now possible to build very fast, low-cost microprocessors. The speed of the current day microprocessors is within an order of magnitude of the speed of the fastest serial processor on the market. Since the cost of microprocessors is very low, it is possible to utilize a large number of them to build low-cost, high-performance systems.

There are applications that can be partitioned to run on architectures that utilize thousands of inexpensive microcomputers (work stations or personal computers) interconnected by a network. For instance, consider a VLSI design environment in which, each engineer performs his part of the design at his workstation, and accesses the shared database and design information (sparingly) over the *network*. Each workstation provides the speed needed for the engineer's tasks, while the communication speeds of the network are adequate for the interaction requirements.

There are applications in which the communication speed of the above network architecture becomes a bottleneck. That is, the processors need to be more closely coupled to attain a very high communication speed. Such performance is expected to be achieved by *Massively Parallel Processor* (MPP) architectures consisting of a large number of processors interconnected by a high bandwidth, low latency network structure.

## I.2 The Need for Advanced Architectures

This section is extracted from the Federal Plan for High-End Computing — Report of the High-End Computing Revitalization Task Force, May 2004. Several terms and concepts mentioned in this section would not be meaningful for the first time reader. Most of these concepts are described later in this book.

In the past decade, computer modeling and simulation of physical phenomena and engineered systems have become widely recognized as the "third pillar" of science and technology — sharing equal billing with theory and experiment. Simulations are performed on computing platforms ranging from simple workstations to very large and powerful systems known as high-end computers (also called supercomputers). High-end computers enable investigations heretofore impossible, which in turn have enabled scientific and technological advances of vast breadth and depth. High-end computing (HEC), thus, has become an indispensable tool for carrying out important missions in science and technology.

Complex systems such as aircraft, proteins, human organs, nuclear weapons, the atmosphere, and stars can be analyzed and better understood through computer models. With advances in high-end computing power, scientists will be able to model such systems in far greater detail and complexity, and eventually to couple individual models to understand the behavior of an entire system. The opportunity for accelerating progress in many fundamental and applied sciences is compelling.

In view of these opportunities, the Office of Science and Technology Policy (OSTP) determined that an effort focused on high-end computing was warranted. The High-End Computing Revitalization Task Force (HECRTF) was chartered under the National Science and Technology Council (NSTC) to develop a plan for undertaking and sustaining a robust Federal high-end computing program to maintain U.S. leadership in science and technology.

The HECRTF solicited input from leading applications scientists in a variety of disciplines who use high-end computing to advance their research. They were asked to identify important scientific challenges addressable by high-end computing and to estimate the additional computational capability (as a multiple of present high-end capability) needed to achieve the goal. The estimates of additional capability needed to achieve the goals ranged from 100 to 1000 times the capability of today's high-end computing resources. For example, fundamental understanding of the emergence of new behaviors and processes in nanomaterials, nanostructures, nanodevices, and nanosystems will require a combination of new theory, new design tools, and high-end computing for large-scale simulation. Similarly, our ability to provide accurate projections of regional climate requires ensembles of simulations on high-end computers at ultrahigh resolution, with sophisticated treatment of cloud formation and dispersal, atmospheric chemistry, and regional influences. The intelligent community's capability to safeguard our national hinges is in part on the ability of high-end computing to tackle diverse computational applications such as cryptanalysis, image processing of satellite and other data, and signal processing for communications traffic, radar, and other signals.

The fastest supercomputers of today are capable of sustaining about 36 trillion calculations per second (Japan's Earth simulator built by NEC in 2002). There are some supercomputers that have reached many times that speed, but not on a sustained basis. IBM's Blue Gen/L, being assembled for Lawrence Livermore National Laboratory, performed 135.3 trillion floating point operations per second running benchmark software. But such speed is not adequate for many scientific applications. The geophysicists for example, need a computer system that is 1000 times faster than today's supercomputers for reliable climate modeling. In a project at the Oak Ridge National Laboratory, scientists envision a computer capable of sustaining 50 trillion calculations per second.

The Report of the HECRTF highlights the following requirements in hardware, software, and systems technologies.

## I.2.1 Hardware

### I.2.1.1 Microarchitecture

The most recent version of the International Technology Roadmap for Semiconductors predicts that microprocessors with 1.5 billion transistors will be introduced by 2010. However, common business economics will continue to drive microarchitecture design to optimize performance for high-volume commercial applications. Optimizing chip functionality to support HEC performance requires different chip architectures and subsystems that may not have immediate impact in the commercial market. The challenge lies in utilizing rich transistor budgets of the future in innovative designs to more effectively address the requirements of high-end applications. Opportunities for microarchitecture research include dynamic optimization of hardware, latency-tolerant mechanisms, introspection, fault isolation and recovery, as well as several others. In addition, nontraditional and reconfigurable processors (e.g., architectures based on Field Programmable Gate Arrays (FPGA), Processor in Memory (PIM), or Application Specific Integrated Circuits (ASIC)) promise substantial improvements in performance, cost, and thermal generation for many HEC applications.

### I.2.1.2 Memory technologies

The clock rates of commodity microprocessors have risen steadily for the past 15 years, while memory latency and bandwidth have improved much more slowly. For example, Intel processor clock rates have doubled roughly every 2 years, from 25 MHz in 1989 to over 3 GHz today as a result of Moore's Law. Memory latency has improved at only 7% per year. Thus a rule of thumb for HEC systems now becomes "Flops are free, but bytes are expensive." As a result, HEC application performance often falls well below 10% of peak, partly because processors stall while waiting for data to operate on.

Memory latency is not perceived as a problem for consumer applications, which typically spend most of their time waiting for human response. It is also not a large problem for large commercial applications, which can easily divide the many simultaneous processes across additional nodes requiring little communication. Attacking the processor–memory bottleneck for large-data, high communication, or random access HEC applications will require research and development (R&D). This area of research is expected to overlap with investigations into microarchitectures and nontraditional processors because of continued increases in the scale of integration afforded by the advances in silicon processing technology. Promising areas of R&D include design of memory subsystems that support latency-tolerant processors, such as morphable processors, tiled architectures, processors with larger caches, faster memories, fast memory I/O drivers and buses, intelligent or adaptable caches, and smart memories.

### *1.2.1.3 Interconnect technologies*

The processor–memory bottleneck is only part of the communication challenge in HEC systems. Communication latency and bandwidth limits must be greatly improved across the entire hierarchy of interconnects: intra-chip, chip-to-chip, board-to-board, cabinet-to-cabinet, and system-to-peripheral (e.g., storage and network). R&D is needed to develop new interconnect technologies (such as optical-based, high-end Infiniband) and architectures (fabrics) that scale to the largest HEC systems, with much lower latency and much higher bandwidth than are commercially available.

### *1.2.1.4 Packaging and thermal management*

Modern processors and other integrated circuits (ICs) are driving advancements in commercial packaging and thermal management technologies. HEC systems drive these technologies even further, with requirements for the most aggressive integration (to minimize communication latency) and thermal dissipation (to remove heat generated by more tightly packed and higher-frequency processors and ICs). The July 2002 *Report on High Performance Computing for the National Security Community* lists a number of IC packaging and thermal management technologies that require R&D to address the demands of high-end computing. Packaging technologies include multichip modules, stacked IC, wafer scale integration, new board and module technologies, new assembly technologies, and others. Thermal management of future HEC systems will require R&D for new liquid, air, and cryogenic cooling technologies beyond what the market will produce on its own.

### *1.2.1.5 Input/Output and storage*

Scalable high-performance I/O mechanisms, file systems, and storage devices are essential to high-end computing. Terascale and petascale applications generate massive amounts of data that must be stored, managed, and assimilated. Storage technology trends are resulting in significant increases in storage capacity, but bandwidth and latency improvements occur much more slowly. Major improvements in scalability throughout I/O and storage are required to keep pace with data generation capability and storage capacity. Otherwise, the risk of write-once, read never data is substantial.

### I.2.2 Software

#### I.2.2.1 Operating systems

A key roadblock to the realization of HEC systems with high sustained performance is the extreme difficulty of scaling commercial operating systems to effectively utilize large numbers of processors (greater than 64) as a single system. Most commercial operating systems cannot maintain cache coherency and a single-system image (SSI) even at this modest scale, and future HEC systems will contain 100,000 or more processors. R&D is needed to scale operating systems to support much larger SSI. This would allow application developers to use relatively painless parallel programming models (e.g., Open specification for Multiprocessing, OpenMP and memory-level parallelism, MLP) as compared to message-passing interface (MPI), and it would push system vendors to develop very fast and efficient interconnect fabrics, which would help address the memory bottleneck as well. Operating systems should also address reliability, availability, and serviceability (RAS), automatically handling software and hardware faults. In general, run-time capabilities also need substantial enhancement to support high-end computing, not just to address RAS, but also for automated data movement (e.g., for distributed shared memory, DSM), load balancing, and scheduling. Software libraries, languages, and compilers all need special attention in the high-end computing context.

#### I.2.2.2 Compilers

HEC vendors supply a base set of compilers (e.g., FORTRAN, C, C++) with a HEC system. However, these compilers are developed and optimized for commercial systems and applications, and are typically not well optimized for the HEC systems of interest to the scientific and engineering communities. Furthermore, there is usually no support for languages that support the DSM programming model, such as Unified Parallel C or Co-array Fortran. The July 2002 *Report on High Performance Computing for the National Security Community* describes the difficulty of building wide vendor support and user adoption of languages that are preferable for high-end computing. Thus, R&D beyond what vendors will undertake is needed for HEC compilers to maximize sustained and parallel application performance to the full size of the system. These compilers must be fully aware of the HEC system processor, memory, and interconnect architectures. In the long term, high-level languages, compilers, and libraries for performance portability across an array of HEC systems, including reconfigurable and nontraditional processor-based systems, are also HEC challenges with little commercial value, but with a potentially major benefit to HEC time to solution.

### 1.2.2.3  Programmer tools and environments

The current class of HEC systems is extremely difficult to program for or port applications to, such that without a heroic effort, applications rarely achieve more than a few percent of the peak capability of the system. The development of better programming tools (porting, debugging, scaling, and optimization) would substantially improve this situation, enabling true performance portability to new architectures. These tools must become vastly easier to use, totally seamless (integrated into an Intelligent Development Environment, IDE), completely cross-platform, and highly efficient when applied to applications running at the full-scale of the system. Industry representatives have stated that the development of such a totally seamless environment is well outside their scope and will take additional R&D by the high-end community. Promising technologies include tools that are visual, language-free, highly automated and intelligent, and user-friendly, and that address more than just CPU performance. Such high-level programming tools should eventually include support for nontraditional HEC systems, for example, based on reconfigurable FPGA processors or PIMs.

### 1.2.2.4  Algorithms

Algorithmic improvements can yield performance improvements at least as significant as those realized by hardware and architectural advances. For the past decade or more, algorithmic research has focused intensely on the capabilities of a single class of architecture — clustered symmetric multiprocessing (SMP) systems. Emerging architectures will be radically different either in scale of implementation or in fundamental computational characteristics that must be taken into account in the development of next-generation algorithms. It is imperative that research in algorithms that can better exploit the current architectural solutions as well as future approaches be conducted in parallel with hardware, software, and systems research.

## 1.2.3  Systems

### 1.2.3.1  System architectures

Recent HEC investments in the United States have focused on cluster-based systems almost exclusively. These cluster-based architectures are not well suited for all applications. Indeed, there are national priority applications that might better be mapped onto systems having very different architectural features (e.g., significantly improved communication to computation ratios). The current HEC focus on clustering hundreds of small nodes, each with a separate operating system, results in poor parallel efficiency, generally below 10% and sometimes lower than 1% of peak on some applications. There is a need for processors matched to interconnect fabrics that perform well on global data access. Larger shared-memory nodes can dramatically reduce programming effort and increase parallel efficiency. The ability to deliver sustained performance on HEC systems will depend to a large degree on the successful achievement of data choreography. This is another reflection of the reality that data movement is becoming more of a limiting factor than arithmetic operations for sustained performance in HEC systems. Successful system architectures will feature high-bandwidth interconnects with minimal latency, tightly integrated network interfaces, large numbers of outstanding memory references, and control over data placement and routing. Effective node architectures (e.g., larger shared-memory/single-system-image nodes) are needed to facilitate scalability and ease of use.

### I.2.3.2  *System modeling and performance analysis*

There is a need for system/architecture modeling and simulation to allow users to evaluate potential configurations against application requirements and to determine properly "balanced systems" for their applications. The systems being developed today and those required to satisfy petaflop needs will be so complicated that simulation and modeling of these systems, at various levels of detail, will be required to explore technological alternatives. These models could also be used to establish a system performance signature that could subsequently be used in assessing a system's viability in the context of application requirements. Finally, these system models could be used to explore parallel programming paradigms and models of I/O performance.

### I.2.3.3  *Reliability, availability, and serviceability and security*

Substantial increases in size and complexity are expected as systems grow to the petascale level and beyond. It is entirely likely that the rate of parts failure will become high enough relative to application execution time that the occurrence of faults during execution will be the norm rather than the exception. This will most likely require an entirely new approach to system reliability and fault management. Focused high-end computing research in RAS, fault tolerance, and system self-awareness is necessary. This research should be closely coupled with system simulation and modeling to facilitate the development of new approaches and yield deep understanding of petascale RAS issues. Scalable security mechanisms are essential to maintaining the integrity of HEC systems, and the complexity of new architectures will have to be accounted for in focused security research.

### I.2.3.4  *Programming models*

Alternative programming paradigms to MPI (or MPI-like) are needed, including methods for evaluating their efficacy, as are methods to accelerate collective operations, perform remote direct memory access (DMA), and offload work to a Network Interface Card (NIC). Advances in new programming models must complement advances in new architectural approaches. Systems consisting of hundreds of thousands of processors must be abstracted if any productivity improvements are to be realized. This effort also requires that sufficient attention be paid to properly socializing the concepts and ideas to the entire high-end computing community.

While the needs vary substantially from application to application, it is clear that all science and engineering domains would benefit from significantly larger, scalable computer architectures that exhibit an improved balance in processor performance, memory bandwidth/latency, communications bandwidth, and improved programming environments. However, simply increasing the size of massively parallel architectures based on current off-the-shelf (COTS) technologies will not effectively address the capability needs of science and engineering research in the coming decade. Many applications have a critical requirement for high-end computers with high-random access global memory bandwidth and low-latency interprocessor communications. While such applications are well suited to symmetric multiprocessors, they are not readily programmable on commodity clusters. Scalable MPP and cluster systems, while providing massive amounts of memory, are inherently more difficult to program. Even emerging parallel vector architectures, such as the Cray SV2/X1, will also be more difficult to program than previous parallel vector systems because of their nonuniform memory access rate.

Numerous attempts are currently under way to retool codes in application areas such as stockpile stewardship, global climate modeling, computational fluid dynamics, local and regional weather forecasting, aircraft design, cosmology, biomolecular simulation, materials science, and quantum chemistry to run more efficiently on MPP architectures, simply because they are the most plentiful systems currently available for high-end computing. While these efforts have resulted in more scalable codes in the short run, they have diverted attention away from the development of systems that provide high-bandwidth access to extremely large global memories. While some of these research areas stand to benefit significantly from factors of 10 to 100 increases in MPP scale, in general they will still not be able to capture more than about 3 to 20% of the peak processor performance on these systems.

Simulations in astrophysics, climate modeling, computational biology, high-energy physics, combustion modeling, materials research, and other scientific applications will routinely produce datasets over the next few years that range in size from hundreds of terabytes to tens of petabytes. It is estimated that the high-energy and nuclear physics community will have several large-scale facilities by 2006 that will generate tens of petabytes of data per year. The full promise of high-end computing will not be realized without adequate storage, file systems, and software to manage, move, and analyze these datasets effectively. A well-balanced HEC facility demands a smooth flow of data to appropriately large storage and file systems and the technology to efficiently search and mine the data during subsequent analyses. Close

# Introduction

coupling between scientific data management and other technologies, such as metadata cataloging, data mining, knowledge extraction, visualization, and networking, must be improved. Current technologies for transferring data from MPP systems to external disk or tape systems will be inadequate by factors of 5 to 50, and our current ability to move even terabytes of data over a network for analysis is severely limited by existing communications bandwidth. Significant investments must be made in storage infrastructure and storage technology over the next several years to deal with the tsunami of scientific data that will be produced.

## I.3 Book Overview

The book follows the progression in architecture technology which started with uniprocessor architectures. Enhancements were made to the uniprocessor architecture to improve its performance. Pipelined architectures were introduced as a natural way of enhancing performance by utilizing the overlapped processing mode. Architectures up to this point basically adopted the serial processing paradigm. The 1990s saw the introduction of parallel architectures that attempt to exploit the parallelism in application algorithms, in order to meet the performance requirements. The book can be viewed as consisting of four parts:

PART 1: Preliminaries (Chapters 1 and 2)

PART 2: Pipelined Architectures (Chapters 3 and 4)

PART 3: Parallel Architectures (Chapters 5 and 6)

PART 4: Current Directions in Architecture (Chapter 7).

Chapter 1 briefly traces the evolution of digital computer architectures starting from the von Neumann single processor structure and details the features of contemporary single-processor systems.

Chapter 2 introduces the most basic models for serial, pipelined, and parallel processing and the corresponding terminology. Along with the evolution of architectures, several architectural classification schemes have also been attempted over the years. A popular classification scheme is utilized in this chapter as a means of introducing styles of architecture.

As mentioned earlier, the first effort towards parallel processing was through pipeline architectures, although such architectures offer more of an overlapped processing mode rather than a strict parallel-processing mode. Pipelined structures were viewed as speed enhancement schemes for single processor architectures. Chapter 3 covers pipeline processing.

Pipeline concepts are extensively used in creating architectures suitable for array- or vector-oriented computations. These architectures, known as *pipelined array processors* or *vector processors*, are covered in Chapter 4.

Chapter 5 describes *synchronous array processor* architectures. These architectures are based on parallelism in data and hence are suitable for computations with array-oriented data. There are several common interconnection structures used to connect the subsystems of a multiple processor system. Interconnection structures suitable for array processor architectures are also described in Chapter 5.

Chapter 6 covers multiprocessor architectures. These are the most generalized parallel-processing structures.

Chapter 7 concludes the book with some of the current architectural trends to include Dataflow computing, Grid computing, Biology-inspired computing, and Optical computing.

The field of computer architecture is a dynamic one. The characteristics of architectures not only keep changing, but architectures fade away and new ones appear at a very fast rate. As such, the example architectures in this book are current as of this writing and the reader is referred to the manufacturer's manuals for the latest details. Commercially available architectures have been selected as examples throughout the book, as far as possible. The book does not include the details on several prominent experimental architectures, since these experiments tend to be special-purpose in nature and may or may not result in commercial production. The reader is referred to magazines such as IEEE Computer, IEEE Spectrum, Computer Design, and Proceedings of the Symposium on Computer Architecture held annually, for details on these architectures.

## References

*Federal Plan for High-End Computing — Report of the High-End Computing Revitalization Task Force*, May 2004.

U.S. Department of Defense, *Report on High-Performance Computing for the National Security Community*, July 2002.

# 1
## Uniprocessor Architecture Overview

The pipelined- and parallel-processing architectures utilize multiple processors and can be considered as enhancements of modern day uniprocessor (or single-processor) architectures. All commercially successful uniprocessor architectures are based on the digital computer model proposed by von Neumann in the 1940s (Burks et al., 1946). There have been several enhancements to this model. Readers not familiar with the von Neumann model and the enhancements thereof are referred to Shiva (2000). In order to provide a baseline for the modern day uniprocessor architecture, a brief review of the von Neumann uniprocessor model is provided in Section 1.1 and a listing of advances in architecture as enhancements to this model are given in Section 1.2. Section 1.3 provides a brief description of two contemporary architectures.

### 1.1 Uniprocessor Model

Figure 1.1 shows the von Neumann model, a typical uniprocessor computer system consisting of the memory unit, the arithmetic logic unit (ALU), the control unit, and the input/output (I/O) unit. The memory unit is a single port device consisting of a memory address register (MAR) and a memory buffer register (MBR) — also called a memory data register (MDR). The memory cells are arranged in the form of several memory words, where each word is the unit of data that can be read or written. All the read and write operations on the memory utilize the memory port. The ALU performs the arithmetic and logic operations on the data items in the accumulator (ACC) and MBR and typically the ACC retains the results of such operations. The control unit consists of a program counter (PC) that contains the address of the instruction to be fetched and an instruction register (IR) into which the instructions are fetched from the memory for execution. A set of index registers are included in the structure. These allow easier access to subsequent locations in the memory. They can also be used to expand the direct addressing range of the processor. For simplicity, the I/O subsystem is shown to input to and output from the ALU subsystem. In practice, the I/O may also occur directly between the memory and I/O devices without utilizing any processor registers. The components

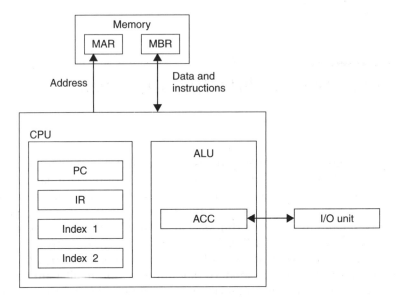

**FIGURE 1.1**
von Neumann architecture

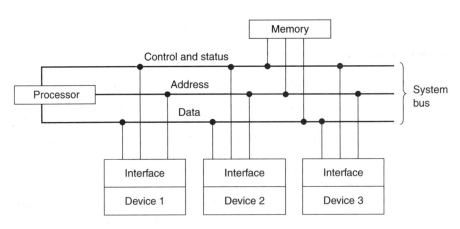

**FIGURE 1.2**
General computer system structure

of the system are interconnected by a multiple-bus structure on which the data and addresses flow. The control unit manages this flow through the use of appropriate control signals.

Figure 1.2 shows a more generalized computer system structure representative of modern day architectures. The processor subsystem (i.e., the central processing unit, CPU) now consists of the ALU and control unit and various processor registers. The processor, memory, and I/O subsystems are interconnected by the system bus which consists of data, address, and control and status lines.

Practical systems may differ from the single-bus architecture of Figure 1.2, in the sense that they may be configured around multiple buses. For instance, there may be a memory bus that connects the processor to the memory subsystem and an I/O bus to interface I/O devices to the processor, forming a two-bus structure. Further, it is possible to configure a system with several I/O buses wherein each bus may interface one type of I/O device to the processor. Since multiple-bus structures allow simultaneous operations on the buses, a higher throughput is possible, compared to single-bus architectures. But, because of the multiple buses the system complexity increases. Thus, a speed/cost tradeoff is required to decide on the system structure.

It is important to note the following characteristics of the von Neumann model that make it inefficient:

1. Programs and data are stored in a single sequential memory, which can create a memory access "bottleneck."
2. There is no explicit distinction between data and instruction representations in the memory. This distinction has to be brought about by the CPU during the execution of programs.
3. High-level language (HLL) programming environments utilize several data structures (such as single and multidimensional arrays, linked lists, etc.). The memory, being one-dimensional, requires that such data structures be linearized for representation.
4. The data representation does not retain any information on the type of data. For instance, there is nothing to distinguish a set of bits representing floating-point (FP) data from that representing a character string. Such distinction has to be brought about by the program logic.

Because of the above characteristics the von Neumann model is overly general and requires excessive mapping by compilers to generate the code executable by the hardware from the programs written in HLLs. This problem is termed as the *semantic* gap. In spite of these deficiencies, the von Neumann model has been the most practical structure for digital computers. Several efficient compilers have been developed over the years that have narrowed the semantic gap to the extent that it is almost invisible for a HLL programming environment.

Early enhancements to the von Neumann model were mainly concentrated on increasing the speed of the basic hardware structure. As the hardware technology progressed, efforts were made to incorporate as many HLL features as possible into firmware and hardware in an effort to reduce the semantic gap.

Note that the hardware enhancements alone may not be sufficient to attain the desired performance. The architecture of the overall computing environment starting from the algorithm development to execution of programs needs to be analyzed to arrive at the appropriate hardware, software, and firmware structures. If possible, these structures should exploit the parallelism in the algorithms themselves. Thus, performance enhancement is one reason for parallel processing. There are also other reasons such as reliability, fault tolerance, expandability, modular development, etc. that dictate parallel-processing structures. Chapter 2 introduces parallel-processing concepts further. The hardware enhancements to von Neumann model are traced briefly in the next section.

## 1.2 Enhancements to the Uniprocessor Model

Note that the von Neumann model of Figure 1.1 provides one path for addresses and a second path for data and instructions, between the CPU and the memory. An early variation of this model is the *Harvard architecture* shown in Figure 1.3. This architecture provides independent paths for data addresses, data, instruction addresses, and instructions. This allows the CPU to access instruction and data simultaneously. The name Harvard *architecture* is due to Howard Aiken's work on Mark-I through Mark-IV computers at Harvard university. These machines had separate storage for data and instructions. Current Harvard architectures do not use separate storage for data instructions but have separate paths and buffers to access data and instructions simultaneously.

The remainder of this section highlights the major performance attributes of each subsystem in the uniprocessor model. A brief discussion of the important

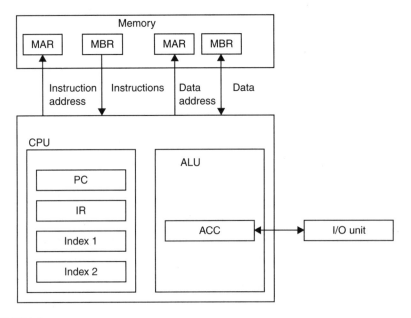

**FIGURE 1.3**
Harvard architecture

architectural features found in practical computer systems, which enhance one or more of the performance attributes is also provided.

### 1.2.1 Arithmetic Logic Unit

The major performance parameters for an ALU are:

1. The variety of arithmetic and logic functions it can perform (i.e., the functionality)
2. The speed of operation

In early 1960s when hardware costs were high, a minimum set of arithmetic (add, subtract) and logic (shift) operations were implemented in hardware. The other required operations were implemented in software. As the progress in integrated circuit (IC) technology yielded cost-effective and compact hardware, several fast algorithms were devised and implemented in hardware. Multifunction ALUs, fast multipliers, and other ICs that aid the design of versatile ALUs are now available off-the-shelf. The ALU subsystem has thus been enhanced in the following ways:

1. Faster algorithms for ALU operations and corresponding hardware implementations (carry lookahead adders, fast multipliers, hardware division, etc.).
2. Use of a large number of general-purpose registers in the processor structure, wherein most of the ALU operations are on the data items residing in those registers. This reduces the number of accesses to the memory and hence increases the speed of the ALU.
3. Stack-based ALUs increased the speed of operation since operands and intermediate results could be maintained in stack registers, thus reducing the memory access requirements. Note that this concept refers to the so called *zero-address machines*, in which the arithmetic and logic operations always refer to the operands on the top one or two levels of a stack. A set of registers in the processor are interconnected to form a stack.
4. The speed of ALU was further enhanced by implementing it as a pipeline with several stages, when the ALU operations allow such multiple stage execution. Example I.1 introduced the concept of the pipeline. Chapter 3 describes this concept further.
5. Implementations of the processing unit with multiple functional units, where each unit is dedicated to a single arithmetic or logic operation (such as add, shift, multiply, etc.) were used in machines built as early as the 1960s (e.g., Control Data Corporation 6600). This implementation allows the simultaneous execution of several functions thereby enhancing the processing speed.

6. As the hardware became more cost-effective, systems with multiple ALUs became common. Here each ALU is capable of performing all required arithmetic/logic functions and all the ALUs operate simultaneously. The array processor attachments and the arithmetic coprocessors for various microprocessors available today are examples of these architectures.

### 1.2.2 Memory

The major parameters of interest for the memory subsystem are:

1. The access speed
2. The capacity
3. The cost

In general, the memory subsystem should provide the highest access speed and the largest capacity at the lowest cost per bit of storage. The speed is measured in terms of two parameters: the bandwidth and the latency. The *bandwidth* is a measure of the number of data units that can be accessed per second. It is a function of the *access time* (i.e., time between a read request and the arrival of data at the memory output) and the cycle time (i.e., the minimum time between requests to memory). The memory *latency* is a measure of how fast the requested data arrives at the destination. It is a function of the bandwidth of the memory, the memory hierarchy, and the speed of the interconnection structure between the memory and the device accessing the memory. Other parameters such as power consumption, weight, volume, reliability, and error detection/correction capability are also important depending on the application.

Several hardware technologies with varying speed-to-cost characteristics have been used in building the memory subsystem. Magnetic core memories were used extensively as the main memory of the systems built through the late 1960s. They are still being used in some space and defense applications where their nonvolatile characteristic is important. But, semiconductor memories have now displaced them almost completely and offer better speeds and lower cost. Static memory devices offer higher speeds compared to dynamic memories, while dynamic memories offer higher chip densities. Access speeds and chip capacities of semiconductor memories have continued to increase due to progress in IC technology.

The following architectural features have been used over the years to enhance the bandwidth and capacity of memory subsystems while maintaining the lower cost:

1. Wider word fetch
2. Blocking (Interleaved and Banked organizations)

3. Instruction/data buffers
4. Cache memories
5. Virtual memory schemes
6. Multiport memories

The memory subsystem is usually built out of several physical modules (or blocks). The size of the module is dependent on the technology and the sizes offered by that technology, as illustrated by the following example:

---

**Example 1.1**
If a 128K words (1K = $2^{10}$ = 1024) memory with 32 bits/word (i.e., 128K × 32 memory) is required and the technology offers (1K × 8) chips, several memory organizations are possible. Some of these are shown below:

| Number of chips per block | Block capacity | Number of blocks |
|---|---|---|
| 4  | 1K × 32  | 128 |
| 8  | 2K × 32  | 64  |
| 16 | 4K × 32  | 32  |
| 64 | 16K × 32 | 8   |

Typically, each block would have its own 'port' (i.e., MAR/MBR) and read/write control circuits and can be accessed independently. As such, it is possible to access several modules simultaneously.

---

One of the earliest of the speed enhancement techniques used in memory systems was the *wider word fetch*. Here, several memory blocks utilize a common MAR and each block has its own MBR. This organization allows access to multiple memory words per memory cycle. Obviously, this requires external buffering of the multiple words accessed from the memory.

Another speed enhancement technique is memory *interleaving* which distributes the memory addresses such that concurrent accesses to memory blocks are possible. In low-*order* interleaving, the blocks are organized such that the consecutive memory addresses lie in consecutive physical blocks, with each block having its own port and read/write controls. That is, if the memory contains $2^n$ words arranged in $2^m$ blocks, the low-order $m$ bits of the address select a block and the remaining $n - m$ bits select a word within that block. This organization allows overlapped access to successive memory locations, since the read/write access to the block containing a word can be initiated well before the access to previous block is completed. This increases the speed of instruction fetch during program execution.

In *high-order interleaving* (also called *banking*) technique, the consecutive addresses lie in the same physical bank. That is, the high-order $m$ bits of the address select a physical block and the remaining $n - m$ bits select a word within that block. This allows the data and instruction segments of the program be located in different physical blocks which can be accessed simultaneously, since each block has its own port and read/write control.

When any of the above speed enhancement techniques are used, a buffer is needed between the memory and the processor to hold the instructions or data. These buffers can take the form of a simple queue where the memory feeds the data or instructions from one end and the processor retrieves it at the other end. It is possible to allow the input to and output from the buffer to occur in parallel as long as proper controls are exercised.

The *cache* memory scheme can be considered as an extension of the idea of memory buffers where a fast (typically, 10 to 100 times faster than the main memory) memory block is inserted between the primary memory and the processor. The motivation for the cache memory was not to buffer data between the processor and the memory, but to take advantage of temporal locality in memory references during program execution, keeping those blocks of memory referenced most recently in the cache, thus decreasing the effective access time. That is, the processor accesses instructions and data from the cache as far as possible. When the cache does not contain the needed data or instruction, an appropriate main memory block is brought into the cache. Almost all modern architectures utilize cache memory mechanisms. Some have separate data and instruction caches, some utilize cache only for instruction retrieval, and some use a single cache for both data and instruction retrieval. Most of the modern single-chip processors have on-chip cache structures. In addition, it is possible to include an off-chip (second level) cache. Set associativity refers to the way the cache memory is organized. When the processor requests the contents of a memory location, only a small set of locations are searched for a match to the given address. As shown in Figure 1.4, a set-associative cache can be thought of as having several ways that comprise a series of sets that are all a cache-line in length. Therefore, there are two ways in a two-way set-associative cache (Shiva, 2000).

*Virtual memory* is used to increase the apparent capacity of the primary memory. Here, a large low-cost secondary memory device (usually magnetic or optical disks or low-speed semiconductor memories) is included in the memory system. The operation of virtual memory system is similar to that between the cache and the main memory. Only the immediately needed blocks of the secondary memory are brought into the primary memory for execution. The blocks in the primary memory (i.e., pages) are replaced from those from the secondary memory as and when needed. This organization allows the programmer to assume that the system memory is as large as the secondary memory. The virtual memory operation is transparent to the programmer.

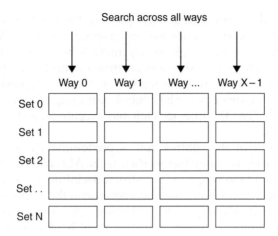

**FIGURE 1.4**
Organization of an X-way set-associative cache

Modern memory architectures are organized in a hierarchy consisting of processor registers, primary cache, main memory, secondary cache (disk cache), and secondary memory. The lowest cost (per bit) and speed characteristics are offered by the secondary memory and the highest cost (per bit) and speed are offered by processor registers. Memory architecture thus involves the selection of appropriate devices and capacities at each level in the hierarchy, to minimize the cost and maximize the speed.

*Multiport* memory ICs are now available. These devices contain multiple ports and individual port controls that allow simultaneous access to data stored in the memory. That is, one port can be reading from a memory location while the other could be writing into another location or several ports can be reading from the same location. But, when multiple ports try to write to the same location, only one of the ports (i.e., the one with the highest priority among them) will be successful.

As described later in this book, multiport memories are useful in building multiple processor systems since the memory subsystem resolves the memory contention problem brought about by the multiplicity of processors accessing the memory simultaneously. Note that large capacity multiport memories are not generally a viable option because of the packaging costs and pin-bandwidth limitations.

### 1.2.3 Control Unit

The major parameters of interest for control units are:

1. Speed
2. Complexity (cost)
3. Flexibility

The design of the control unit must provide for the fastest execution of instructions possible. Instruction execution speed obviously depends on the number of data paths (i.e., single versus multiple buses) available in the processor structure, the complexity of the instruction itself (number of memory addresses, number of addressing modes, etc.) and the speeds of hardware components in the processor. The control unit design should minimize the instruction cycle (i.e., the time to fetch and execute an instruction) time for each instruction.

The complexity of the control unit is predominantly a function of the instruction-set size, although factors such as the ALU complexity, the register set complexity, and the processor bus structures influences it. Machines built in early 1960s had small instruction sets because of the complexity and hence the high cost of hardware. As IC technology progressed, it became cost-effective to implement complex control units in hardware. Thus, machines with large instruction set (i.e., *complex instruction set computers*, CISC) were built. It was noted that in IC implementation of the CISC, the control unit would occupy 60 to 75% of the total silicon area. It was also observed that on an average 20 to 30% of the instructions in an instruction set are not commonly used by application programmers and it is difficult to design HLL compilers that can utilize a large instruction set. These observations lead to the development of *reduced instruction set computers* (RISC). The RISCs of early 1980s had relatively small instruction sets (i.e., 50 to 100 instructions). But, the instruction sets of the modern day RISCs have 200+ instructions.

The IC technology is currently progressing at such a rapid rate, that newer and more powerful processors are introduced to the market very fast. This forces the computer manufacturers to introduce their enhanced product rapidly enough that the competition does not gain an edge. To make such rapid introduction possible, the design (or enhancement) cycle time for the product must be as short as possible. The control unit, being the most complex of the components of the computer system, consumes the largest part of the design cycle time. Hence, it is important to have a flexible design that can be enhanced very rapidly, with a minimum number of changes to the hardware structure.

Two popular implementations of the control unit are:

1. Hardwired
2. Microprogrammed

Note that each instruction cycle corresponds to a sequence of microoperations (or register transfer operations) brought about by the control unit. These sequences are produced by a set of gates and flip–flops in a hardwired control unit, or from the microprograms located in the control read only memory (ROM) in a microprogrammed control unit. Thus, changing these sequences requires only changing of ROM contents in the case of a microprogrammed control unit, while it requires a redesign of a hardwired control unit.

The microprogrammed control units offer flexibility in terms of tailoring the instruction set for a particular application. The hardwired implementations on the other hand, offer higher speeds.

Almost all hardwired control units are implemented as synchronous units, whose operation is controlled by a clock signal. Synchronous control units are relatively simpler to design compared to asynchronous units. Asynchronous units do not have a controlling clock signal. The completion of one microoperation triggers the next, in these units. If designed properly, asynchronous units provide faster speeds.

A popular scheme for enhancing the speed of execution is the overlapped instruction execution where the control unit is designed as a pipeline consisting of several stages (Fetch, Decode, Address compute, execute, etc.). Chapter 3 provides further details on pipelined control units.

One measure of the performance is the average amount of time it takes a processor to complete a task. The run time $R$ of a task containing $N$ instructions, on a processor that consumes on an average $C$ cycles per instruction, with a clock speed of $T$ seconds per cycle, is given by:

$$R = N \times C \times T \qquad (1.1)$$

Thus, three factors determine the overall time to complete a task. The $N$ is dependent upon the task, the compiler and the skill of the programmer and therefore is not processor architecture dependent. The $C$ and $T$, however, are processor dependent. Reduction of the $T$ is accomplished simply through a high-speed clock. For instance, in the MIPS 4000 processor, the external clock input is 50 MHz. An on-chip phase-locked-loop multiplies this by two to get an internal clock speed of 100 MHz, the speed at which the pipeline runs. This 100 MHz can be divided by 2, 3, or 4 to produce interface speeds of 50, 33, or 25 MHz, thus allowing for some low end of 50 MHz systems. However, the main reason for the adjustable clock speed is to make room for a 75 or 100 MHz external clock (150 or 200 MHz internal clock), while maintaining a 50 MHz system interface.

There are two popular techniques for reducing the $C$: Superpipelining and Superscaling. In *superpipelining*, instructions are processed in long (5- to 10-stage) pipelines. This allows simultaneous (overlapped) processing of multiple instructions, thus enhancing the throughput. *Superscaling* also allows for processing of multiple instructions per clock cycle, but does this through multiple execution units rather than through pipelining. In order for superpipelining to work well, there must be a fast clock. In order for superscaling to work well, there must be good instruction dispatch and scoreboarding mechanisms. Since these are done in hardware, they take up a lot of chip real estate. Superpipelining theoretically lacks scalability, because current technology limits the pipeline at running twice as fast as the cache. Superscaling, however, could theoretically have an unlimited bus size and an unlimited number of execution units, and therefore performance could be improved endlessly. In reality, the multiple dispatch circuitries grow in complexity very rapidly as one increase the dispatch multiplicity past two. Also, the compiler complexity grows very rapidly as one tries to avoid constant stalls with the multiple execution units.

### 1.2.4  I/O Subsystem

The major performance parameter of the I/O subsystem is the speed. This subsystem should provide for the fastest transfer of data between the processing system and the application environment.

The I/O interface needs to be general in the sense that it should be easy to interface newer devices of differing characteristics to the system. Several bus standards and I/O protocols have evolved over the years. The I/O subsystem should accommodate such standards easily.

The most popular I/O structures are:

1. Programmed I/O
2. Interrupt mode I/O
3. Direct memory access
4. Channels
5. I/O processors

In the *programmed* I/O structure, the CPU initiates the I/O operation and waits for either the input device to provide the data or the output device to accept the data. Since I/O devices are slow compared to the CPU and the CPU has to wait for the slow I/O operations to be completed before continuing with other computations, this structure is inefficient. It is possible to increase the efficiency, if the processor is made to go off and perform other tasks if possible while the I/O devices are generating or accepting data. Either way, the processor has to explicitly check for the availability or acceptance of data. This is the simplest of the I/O structures since it requires a simple I/O device controller that can respond to the control commands from the CPU and all the I/O protocol is handled by the CPU. In this structure typically one or more processor registers are involved in data transfer.

In the *interrupt mode* I/O, some of the I/O control is transferred from the CPU to the I/O device controller. The CPU initiates the I/O operation and continues with other tasks (if any) and the I/O device interrupts the CPU once the I/O operation is complete. This I/O mode enhances the throughput of the system since the CPU is not idle during the I/O. This mode of I/O is required in applications such as real-time control where the system has to respond to the changes in inputs which occur at unpredictable times.

The *direct memory access* (DMA) mode of I/O allows I/O devices to transfer data to and from the memory directly. This mode is useful when large volumes of data need to be transferred. In this structure the CPU initiates the DMA controller to perform an I/O operation and continues with its computational tasks. The DMA controller acquires the memory bus as needed and transfers data directly to and from the memory. The CPU and the DMA controllers compete for the memory access through the memory bus. The I/O control is thus transferred completely to the I/O device relieving the CPU from the I/O overhead. Note also that this structure does not require CPU registers to be involved in I/O operations, as in other structures described earlier.

The *I/O channels* are enhanced DMA controllers. In addition to performing data transfer in the DMA mode, they perform other I/O related operations such as error detection and correction, code conversion and control of multiple I/O devices. There are two types of I/O channels. *Selector channels* are used with high-speed devices (such as disk and tape) and *multiplexer channels* are used with slow speed devices (such as terminals). Large computer systems of today typically have several I/O channels.

Traditionally, I/O channels were specially designed hardware elements to suit a particular CPU. With the advent of very large-scale integration (VLSI), low-cost microprocessors came into being. The I/O channels were then implemented using microprocessors, programmed to perform the I/O operations of the channel. As the capabilities of microprocessors were enhanced to make them suitable for general-purpose processing, channels became *I/O processors* that handled some computational tasks on the data in addition to I/O.

The structure of modern day systems thus consists of a multiplicity of processors: a central processor and one or more I/O processors or *front-end processors*. With such multiple processor structure, the computational task is partitioned such that the subtasks are executed on various processors simultaneously (as far as possible). This is a simple form of parallel processing.

### 1.2.5 Interconnection Structures

The subsystems of the computer system are typically interconnected by a bus structure. The function of this interconnection structure is to carry data, address, and control signals. Thus the bus structure can be organized either to consist of separate data, control, and address buses, or as a common bus carrying all the three types of signals or several buses (carrying all the three types of signals) each interconnecting a specific set of subsystems. For instance, in multiple bus structures it is common to see buses designated as memory bus and I/O bus.

The registers and other resources within the CPU are also interconnected by either a single or a multiple-bus structure. But, data transfer on these buses does not involve elaborate protocols as on buses that interconnect system elements (CPU, Memory, I/O devices).

The major performance measure of the bus structure is its *bandwidth*, which is a measure of number of units of data it can transfer per second. The bandwidth is a function of the bus width (i.e., number of bits the bus can carry at a time), the speed of the interface hardware and the bus protocols adopted by the system.

The advantage of multiple-bus structures is the possibility of simultaneous transfers on multiple buses, thereby providing a high overall bandwidth. The disadvantage of this structure is the complexity of hardware.

Single-bus structure provides uniform-interfacing characteristics for all devices connected to it and simplicity of design, because of the single set of interface characteristics. Since this structure utilizes a single transfer path, it can result in higher bottlenecks for transfer operations, compared to multiple-bus structures.

As computer architectures evolved into systems with multiple processors, several other interconnection schemes with varying performance characteristics have been introduced. Several such interconnection structures are described in later chapters of this book.

### 1.2.6 System Considerations

All enhancements outlined in this section were attempts to increase the throughput of the uniprocessor system. As hardware technology moved into the VLSI-era more and more functions were implemented in hardware, rather than in software. Large instruction sets, large number of general-purpose registers, larger memories became common. But the basic structure of the machine remained the one that was proposed by von Neumann.

With the current technology, it is possible to fabricate a complex processing system on a chip. Processors with 64-bit architectures along with a limited amount of memory and I/O interfaces are now fabricated as single chip systems. It is expected that the complexity of systems fabricated on a chip would continue to grow.

With the availability of low-cost microprocessors, the trend in architecture has been to design systems with multiple processors. It is very common to see a CPU and several coprocessors in such multiple processor architectures. The CPU is usually a general-purpose processor dedicated to basic computational functions. The coprocessors tend to be specialized. Some popular coprocessors are: numeric coprocessors that handle floating-point computations working with an integer oriented CPU; I/O processors dedicated to I/O operations and memory management units (MMUs) that coordinate the main memory, cache, and virtual memory interactions.

There are also multiple processor architectures in which each processor is a general CPU (i.e., the CPU/coprocessor distinction does not exist). There are many applications that can be partitioned into subtasks that can be executed in parallel. For such applications the multiple processor structure has been proved to be very cost-effective compared to building systems with a single powerful CPU.

Note that there are two aspects of the multiple processor structures described above that contribute to their enhanced performance. First, each coprocessor is dedicated for a specialized function and as such, it can be optimized to perform that function efficiently. Second, all the processors in the system could be operating simultaneously (as far as the application allows it), thereby providing a higher throughput compared to single processor systems. Due to the second aspect, these multiprocessor systems can be called *parallel-processing* architectures. Thus, trends in technology have forced the implementation of parallel-processing structures. There are other motivations for parallel processing covered later in this book.

## 1.3 Example Systems

This section provides an overview of two commercial architectures: Intel Corporation's Itanium and MIPS computer system's R10000. In an attempt to provide a complete description we have included several advanced architectural features, which may not be meaningful to the first time reader. A revisit to this section is recommended after reviewing the remaining chapters of the book.

### 1.3.1 Intel Corporation's Itanium

The Itanium™ processor (with 733–800 MHz speed) is the first in a family of 64-bit processors from Intel released in 2001, and Itanium™ 2 processor (900 MHz–1 GHz) was launched in 2002.

Itanium™ processor is based on explicit parallel instruction computing (EPIC) technology that provides pipelining and is capable of doing parallel execution of up to six instructions. Some of the EPIC technology features, provided by Itanium™ processor, are:

*Predication.* EPIC uses predicated execution of branches to reduce the branch penalty. Consider the C source code:

```
if (x==7) z = 2
else z = 7;
```

The instruction flow would generally be written as:

1. Compare x to 7
2. If not equal goto line 5
3. z=2
4. goto line 6
5. z=7
6. //Program continues from here

In the above code, line 2 or 4 causes at least one break (goto) in the instruction flow irrespective of the value of x. This causes interruption in the program flow twice in one pass. Itanium architecture assigns the result of the compare operation to a predicate bit, which then allows or disallows the output to be committed to memory. Thus the same C source code could be written using this IA-64 instruction flow:

1. Compare x to 7 and store result in a predicate bit (let it be "P")
2. If P==1; z=2
3. If P==0; z=7

In the above code, if the value of the predicate bit (P) is equal to 1 (which indicates the x==7), z is assigned 2 otherwise 7. Once the P is set, it can be tested again and again in subsequent code.

*Speculation.* Along with predication, EPIC supports control and data speculation to handle branches.

*Control speculation.* Control speculation is the execution of an operation before the branch which guards it. Consider the code sequence:

```
if (a>b) load (ld_addr1, target1)
else load (ld_addr2, target2)
```

If the operation load (ld_addr1, target1) were to be performed prior to the determination of (a>b), then the operation would be control speculative with respect to the controlling condition (a>b). Under normal execution, the operation load (ld_addr1, target1) may or may not execute. If the new control speculative load causes an exception, then the exception should only be serviced if (a>b) is true. When the compiler uses control speculation, it leaves a check operation at the original location. The check verifies whether an exception has occurred and if so it branches to recovery code.

*Data speculation.* This is also known as "advance loading." It is the execution of a memory load prior to a store that preceded it. Consider the code sequence below:

```
store (st_addr, data)
load (ld_addr, target)
use (target)
```

In the above example, if ld_addr and st_addr cannot be disambiguated (the process of determining at compile time the relationship between memory addresses) and if load were to be performed prior to the store, then the load would be data speculative with respect to the store. If memory addresses overlap during the execution, a data-speculative load issued before the store might return a different value than a regular load issued after the store. When the compiler data speculates a load, it leaves a check instruction at the original location of the load. The check verifies whether an overlap has occurred and if so it branches to recovery code.

*Register rotation.* EPIC supports register renaming by using register rotation technique, wherein the name of the register is generated dynamically. The rotating register base (RRB) is present in a rotating register file which is added with register number given in an instruction and modulo the number of registers in rotating register file. This generated number is actually used as register address.

Figure 1.5 shows the block diagram of Itanium processor. The architectural components of the processor include:

1. Functional unit
2. Cache (three levels: L1, L2, and L3)
3. Register stack engine
4. Bus

The processor consists of four floating-point units, four integer units, four multimedia extensions (MMX) units, and an IA-32 decode and control unit. Out of the four floating-point units, two work at 82-bit precision while the other two are used for 32-bit precision operations. All four can perform fused multiply accumulate (FMAC) operations along with two-operand addition and subtraction, as well as floating-point and integer multiplication. It also executes several single-operand instructions, such as conversions between floating-point and integers, precision conversion, negation, and absolute

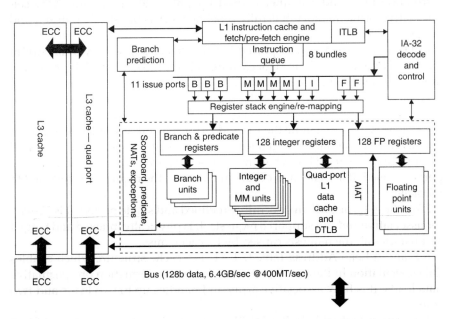

**FIGURE 1.5**
Intel Itanium Architecture (Reproduced from *Intel Itanium 2 Architecture, Hardware Developer's Manual*, July 2002. With permission.)

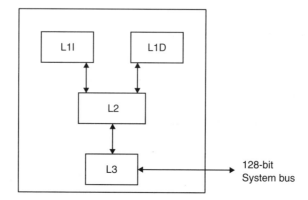

**FIGURE 1.6**
Intel Itanium Cache (Reproduced from *Intel Itanium 2 Architecture, Hardware Developer's Manual*, July 2002. With permission.)

value. Integer units are used for arithmetic and other integer or character manipulations and MMX units to accommodate instructions for multimedia operations. IA-32 decodes and control unit provides compatibility with the other families.

Figure 1.6 shows the cache hierarchy. It consists of three caches: L1, L2, and L3 caches. L1 has separate data (L1D) and instruction cache (L1I).

L1 instruction (L1I) cache is dual-ported 16 KB cache. One port is used for instruction fetch and the other is used for prefetches. It is a four-way set-associative cache memory. It is fully pipelined and is physically indexed and tagged.

L1 Data (L1D) cache is four-ported 16 KB in size. It supports two concurrent loads and two stores. It is used for caching only the integer data. It is physically indexed and tagged for loads and stores.

L2 cache is 256 KB, four ported that supports up to four concurrent accesses. It is eight-way set-associative and is physically indexed and tagged.

L2 cache is used to handle the L1I and L1D cache misses and accesses up to four floating-point data. When data is requested either one, two, three, or all the four ports are used but when instructions are accessed in L2 cache, all the ports are used. If a cache miss occurs in L2 cache, the request is forwarded to L3 cache.

L3 cache is 1.5 or 3 MB in size, fully pipelined and single ported. It is 12-way set-associative and can support 8 requests. It is physically indexed and tagged and handles all the requests caused by L2 cache miss.

Advanced Load Address Table (ALAT) is a cache structure which enables data speculation in Itanium 2 processor. It keeps information on speculative data loads. It is fully associative array and handles up to two loads and two stores.

Translation lookaside buffer (TLB) on the Itanium 2 processor are of two types: data translation lookaside buffer (DTLB) and the instruction lookaside buffer (ITLB). There are two levels of DTLB: L1 DTLB and L2 DTLB. The first level DTLB is fully associative and is used to perform virtual to physical address translations for load transactions that hit in L1 cache. It has three ports: two read ports and one write port. It supports 4 KB pages. The second level DTLB handles virtual to physical address translations for data memory references during stores. It is fully associative and has four ports. It can support page sizes from 4 KB to 4 GB.

The ITLB is categorized into two levels: Level1 ITLB (ITLB1) and Level 2 ITLB (ITLB2). ITLB1 is dual ported, fully associative, responsible for virtual to physical address translations to enable instruction truncation hits in L1I cache. It can support page sizes from 4 KB to 4 GB. ITLB2 is fully associative and responsible for virtual to physical address translations for instruction memory references that miss the ITLB1. It supports page sizes from 4 KB to 4 GB.

Itanium™ contains 128 integer registers, 128 floating-point registers, 64 predicate registers, 8 branch registers, and a 128-register "application store" where machine state is controlled. The Register Stack Engine manages the data in the integer registers.

Figure 1.7 shows the complete register set. It consists of:

- *General-purpose registers*: These are a set of 128 64-bit general-purpose registers, named $gr_0$–$gr_{127}$. These registers are partitioned into static general registers which are 0 through and stacked general registers that are 32 through 127. $gr_0$ always reads 0 when sourced as an operand, and Illegal Operation fault occurs if an attempt to write to $gr_0$ occurs.
- *Floating-point registers*: Floating-point computation uses a set of 128 82-bit floating-point registers, named $fr_0$–$fr_{127}$. These registers are divided into subsets: static floating-point registers that include $fr_0$ through $fr_{31}$ and rotating floating-point registers ($fr_{32}$–$fr_{127}$). $fr_0$ always reads +0.0 when sourced; $fr_1$ always reads +1.0 when sourced. A fault occurs if either of the $fr_0$ or $fr_1$ is used as the destination.
- *Predicate registers*: These are a set of 64 1-bit and are named $pr_0$–$pr_{63}$. They hold the results of compare instructions. These registers are partitioned into subsets: static predicate registers that include $pr_0$–$pr_{15}$ and rotating predicate registers that extend from $pr_{16}$ to $pr_{63}$. $pr_0$ always reads 1 and the result is discarded if it is used as a destination. The rotating registers support software pipeline loops while static predicate registers are used for conditional branching.
- *Branch registers*: It consist of a set of eight 64-bit branch registers, named $br_0$–$br_7$. It is used to hold indirect branching information.
- *Kernel registers*: There are eight 64-bit kernel registers, named $kr_0$–$kr_7$. They are used to communicate information from the kernel (Operating System) to an application.
- *Current frame marker (CFM)*: It describes the state of the general register stack. It is a 64-bit register used for stack-frame operations.

- *Instruction pointer*: It is a 64-bit Pointer and holds pointer to current 16-byte aligned bundle in IA-64 mode, or offset to 1-byte aligned instruction in IA-32 mode.
- *Performance monitor data registers (PMD)*: They are the data registers for performance monitor hardware.
- *User mask* (UM): It is a set of single-bit values used to monitor floating-point register usage, performance monitor.
- *Processor identifiers* (CPUID): They describe processor implementation-dependent features.
- There are several other 64-bit registers with operating system-specific, hardware-specific, or application-specific uses covering hardware control and system configuration.

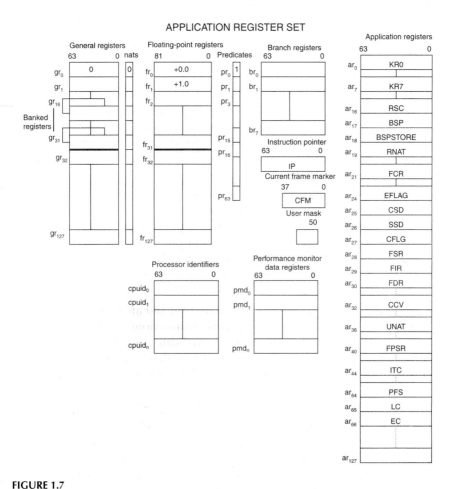

**FIGURE 1.7**
Register set (Reproduced from *Intel Itanium Architecture Software Developer's Manual, Vol. 1: Application Architecture*, October 2002. With permission.)

*Uniprocessor Architecture Overview* 35

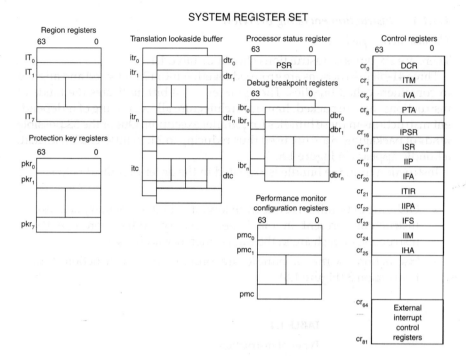

**FIGURE 1.7**
Continued

Register stack engine (RSE) is used to remove the latency (delay) caused by the necessary saving and restoring of data processing registers when entering and leaving a procedure. The stack provides for fast procedure calls by passing arguments in registers as opposed to the stack.

When a procedure is called, a new frame of registers is made available to the called procedure without the need for an explicit save of the caller's registers. The old registers remain in the large on-chip physical register file as long as there is enough physical capacity. When the number of registers needed overflows the available physical capacity, a state machine called the RSE saves the registers to memory to free up the necessary registers needed for the upcoming call. The RSE maintains the illusion of infinite number of registers.

On a call return, the base register is restored to the value that the caller was using to access registers prior to the call. Often a return is encountered even before these registers need to be saved, making it unnecessary to restore them. In cases where the RSE has saved some of the callee's registers, the processor stalls on the return until the RSE can restore the appropriate number of the callee's registers.

The bus in Itanium processor is 128 bits wide and operates at a clock frequency of 400 MHz, totaling to 6.4 GB per second rate.

#### 1.3.1.1 Instruction encoding overview

*The instruction package*

There are six types of instructions shown in Table 1.1.

Three 41-bit instructions are grouped together into 128-bit sized and aligned in containers called bundles. The "pointer" (0–4 bit), indicates the kinds of instructions that are packed. Itanium architecture allows issuing of independent instructions in these bundles for parallel execution. Out of the 32 possible kinds of packaging, 8 are not used thus reducing to 24. In little-endian format, a bundle appears in Figure 1.8.

Instruction from the bundle is executed in the order described below:

- Ordering of bundles is done from lowest to highest memory address. Instructions present in the lower memory address precede the instructions which are in the higher memory addresses.
- Instructions within a bundle are ordered from Instruction 1 to Instruction 3 (Figure 1.8).

**TABLE 1.1**

Types of instructions

| Instruction type | Description |
|---|---|
| A | Integer ALU |
| I | Non-ALU integer |
| M | Memory |
| F | Floating point |
| B | Branch |
| L + X | Extended |

Reproduced from *Intel Itanium Architecture Software Developer's Manual, Vol. 1: Application Architecture*, October 2002. With permission.

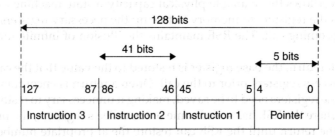

**FIGURE 1.8**
Instruction bundle (Reproduced from *Intel Itanium Architecture Software Developer's Manual, Vol. 1: Application Architecture*, October 2002. With permission.)

# Uniprocessor Architecture Overview

*Instruction set transition model*

There are two operating environments supported by Itanium architecture:

1. *IA-32 System Environment*: supports IA-32, which is 32-bit operating system.
2. *Itanium System Environment*: supports Itanium-based operating systems (IA-64).

The processor can execute either IA-32 or Itanium instructions at any time. Intel Architecture (IA-64) is compatible with the 32-bit software (IA-32). The software can be run in real mode (16 bits), protected mode (32 bits), and virtual mode 86 (16 bits). Thus the CPU will be able to operate in both IA-64 mode and IA-32 mode. There are special instructions to go from one mode to the other, as is shown in Figure 1.9.

There are three instructions and interruptions that make the transition between the IA-32 and Itanium instruction sets (IA-64). They are:

- *JMPE (IA-32 instruction)*: jumps to a 64-bit instruction and changes to IA-64 mode.
- *br.ia (IA-64 instruction)*: moves to a 32-bit instruction and changes to IA-32 mode.
- *rfi (IA-64 instruction)*: it is the return of the interruption; the return happens both to an IA-32 situation and to an IA-64, depending on the situation present at the moment when the interruption is invoked.
- *Interrupts transition*: it is the processor to the Itanium instruction set for all interrupt conditions.

### 1.3.1.2 Memory

Memory is byte addressable and is accessed with 64-bit pointers. Pointers 32 bit are manipulated in 64-bit registers.

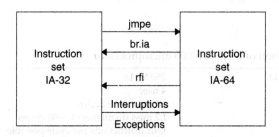

**FIGURE 1.9**
Model of instruction sets transition (Reproduced from *Intel Itanium Architecture, Software Developer's Manual*, December 2001. With permission.)

*Addressable unit and alignment*

Memory in IA-64 can be addressed in units of 1, 2, 4, 8, 10, and 16 bytes. Although data on IA-64 can be aligned on any boundary, IA-64 recommends that items be aligned on naturally aligned boundaries for the object size. For example, words should be aligned on word boundaries. There is one exception: 10-byte floating-point values should be aligned on 16-byte boundaries.

When the quantities are loaded to the general registers from memory, they are placed in the least-significant portion of the register.

*Byte ordering*

There are two endian models: big endian and little endian. Little endian results in the least significant byte of the load operand being stored in the least significant byte of the target operand and the big-endian results in the most significant byte of the load operand being stored in the least significant byte of the target operand. IA-64 specifies which endian model should be used. All IA-32 CPUs are little endian. All IA-64 instruction fetches are performed little endian regardless of current endian mode. The instruction appears in the reverse order if the instruction data is read using big endian.

### 1.3.2 MIPS Computer System's R10000

The MIPS R10000 is a four-way superscalar RISC microprocessor that was introduced in 1996. The four-way superscalar R10000 microprocessor can fetch four instructions and issue up to five instructions per cycle. Some of the specifications of MIPS R10000 are shown in Table 1.2.

Figure 1.10(a) shows the architecture of the MIPS R10000. (Figure 1.10(b) describes the pipelining details, which are described later in this chapter.) The architectural components of the processor include:

1. Functional unit
2. Integer and floating-point register file
3. Integer, floating-point, and address queue

**TABLE 1.2**

Some specification of MIPS R10000 microprocessor

| | |
|---|---|
| Chip speed (clock) | 250 MHz |
| Cycle time | 4 nsec |
| Cache latencies | L1 miss: ~10 cycles |
| | L2 miss (to memory): ~60 cycles |
| | four instructions per cycle possible |
| Translation lookaside buffer (TLB) | TLB holds 64 entries, each of which maps to two consecutive virtual pages. Each pages is 16 KB |
| TLB latencies | 25 cycles if in L2, or more, similar to cache miss |
| Page sizes | 16 KB |

# Uniprocessor Architecture Overview

4. Two levels of cache — on- and off-chip
5. Active list
6. Branch unit
7. Free register list

There are five functional units: two integer ALUs (ALU1 and ALU2), the Load/Store unit which is used for address calculation, the floating-point

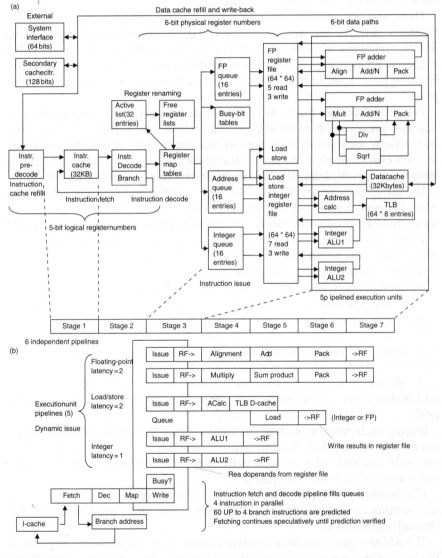

**FIGURE 1.10**
MIPS R10000 Microprocessor architecture (Reproduced from *MIPS R10000, Microprocessor User's Manual*, 1996. With permission.)

adder (FP adder) which is used for additions and the floating-point multiplier (FP multiplier) used for multiplication.

There are also three units to compute more complex results: Integer Multiply/Divide execution unit, Divide execution unit, and Square-root execution unit. Integer Multiply/Divide execution unit performs integer multiply and divide operations. These instructions are issued to Integer ALU2. Divide execution unit (Div) performs the floating-point divides and these instructions are issued to the floating-point multiplier. Square-root execution unit (Sqrt) is used to perform the floating-point square root instructions which are issued to the floating-point multiplier.

There are two register files: integer register file and FP register file. The integer and the FP register file are used to assign registers to the integer and floating-point instructions, respectively. The integer register file consists of 64 registers. It has 7 read and 3 write ports and a separate 64 bit condition file. Floating point register file, on the other hand, consists of 5 read and 3 write ports.

The processor has three instruction queues: integer, floating point, and address queue, for dynamically issuing instructions to the execution units. Instructions are marked with tags so that each queue can track instruction in each execution pipeline stage (described later in the section). When the instruction is completed, the tags in the instruction are set to *Done* bit.

The integer queue contains 16 instruction entries and controls 6 dedicated ports to register file: two operand read ports and a destination write port for each ALU. The instructions in the integer arithmetic units: Integer ALU1 and Integer ALU2 are issued by the integer queue. In each cycle, up to four instructions can be written and the newly decoded integer is written into empty entries in no particular order.

Instruction such as Branch and shift can only be issued to ALU1 whereas Integer multiply and divide instructions can be issued only to ALU2. Other integer instructions can be issued to either ALU.

The floating-point queue is similar to integer queue. It contains 16 instruction entries and issues instructions to the FP multiplier and the FP adder. It controls two operand read ports and a destination port for each execution unit. Multiplier's issue port is used to issue instruction to the square-root and divide units. These instructions also share the FP multiplier's register ports.

During each cycle, up to four instructions may be written and like integer queue newly decoded floating-point instructions are written into empty entries in random order. As soon as the instruction is issued to floating-point execution unit, it leaves the queue.

The floating-point queue uses simple sequencing logic for multiple pass instructions such as Multiply–Add that pass through FP multiplier and then through the FPadder.

The address queue contains 16 instruction entries and issues instructions to the load/store unit. It is similar to the integer queue except that it is organized as a circular first-in first-out (FIFO) buffer which means that the next available

# Uniprocessor Architecture Overview

sequential empty entry stores a newly decoded load/store instruction. Each cycle can write up to four instructions. The FIFO order is useful as it maintains the program's original instruction sequence.

Instructions remain in this queue until they have been issued and cannot be deleted immediately after being issued, since the load/store unit may not be able to complete the operation immediately.

An instruction may fail to complete because of a memory dependency, a sscache miss, or a resource conflict. The address queue must continue to reissue the instruction until it is completed.

The address queue has three issue ports:

- The first port controls two read ports to the integer register file. It issues each instruction once to the address calculation unit. This unit computes and translates instruction's memory address using a two-stage pipeline in the TLB. Address stack stores the address. If the cache is available, the two read ports are accessed at the same time as the TLB.

- Second, instructions can be re-issued to the data cache. The queue allocates four sections of the cache, which consist of the tag and data sections of the two cache banks. Load and store instructions cycle start with a tag check to check if the required address is already in cache. If the address is not already in cache then a refill operation is initiated, and instruction waits until it completes. Load instructions read may be either concurrent to or subsequent to the tag check. The instruction is marked *done* in the queue if the data is present and no dependencies exist.

- Third, the address queue can issue store instructions to the data cache. A store instruction may not modify the data cache until it graduates. Only one store can graduate per cycle, but it may be anywhere within the four oldest instructions, if all previous instructions are already completed.

The access and store ports share four register file ports: integer read and write, floating-point read and write. These shared ports are also used for Jump and Link and Jump Register instructions, and for move instructions between the integer and register files.

The MIPS R10000 uses two levels of cache: on-chip which is the primary level and off-chip which is the secondary level. The on-chip cache has 32 KB for data and 32 KB for instructions and the second level is unified for both data and instructions and ranges from 512 KB to 16 MB. Both the level of cache is set-associative.

Figure 1.11 shows the primary data cache on MIPS R10000. The memory address (Virtual Address) comprises several components: the tag, the index, a bit that determines the cache bank (B), and two bits to determine the specific double word (DW) in a set. The Virtual Address is converted to Physical Page

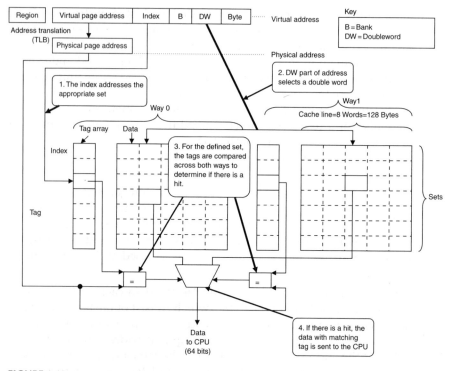

**FIGURE 1.11**
MIPS R10000 primary data cache (Reproduced from MIPS R10000 Microprocessor User's Manual, 1996. With permission.)

Address using the TLB. The cache has a dedicated Tag Array for all the ways that consists of number of entries corresponding to the number of sets. The index appropriately searches the set as soon as the memory address is given to the cache. Once it is known, the Tag Array in each way is searched and a hit occurs if the tag matches the entry. The corresponding two bits in the address then determine DW in a cache line and return the contents of the address to the CPU. In case a match is not found, it is known as cache miss and is reported back to CPU.

Searches in MIPS R10000 are allowed across both ways in parallel that is the tags are checked in both ways simultaneously to share the same path to the processor. It uses a replacement algorithm called as least recently used (LRU) mechanism. The address passes to the secondary cache if the primary cache miss occurs during a read. The data from the secondary cache replaces the least recently used primary cache line if the secondary cache hits. On the other hand if a miss occurs, the least recently used secondary cache line is replaced by the relevant data from main memory.

In MIPS R10000, stored data is always written from the CPU into primary cache instead of writing it directly to memory. This is known as write-back protocol in the cache hierarchy. Write back from the primary data cache goes to the secondary cache and similarly from the secondary cache the write back goes to the main memory. When lines from secondary cache replace lines in primary cache, it is first checked to see if it has been modified. If it has, it is written to secondary cache before being replaced. In a similar way, when a line in secondary cache is replaced it is first checked to see if it has also been modified and, if so, is written back to main memory.

The MIPS R10000 processor's active list records status, such as those instructions that have completed, or those instructions that have detected exceptions. It indicates the physical register which contained the *previous* value of the destination register (if any). It tracks the program-order list of decoded instructions. The previous value is discarded and the physical register is returned to the free list if this instruction graduates. As the instructions graduate, they are removed from top and appended to the bottom of the list as they are decoded. If an instruction causes an exception, active list undoes the results by unmapping the physical register that received the result.

The MIPS R10000 has one branch unit that allows one branch per cycle. It has a 44-bit adder to compute branch addresses. The conditional branches can be executed speculatively up to 4-deep level. It has four-quadword branch-resume buffer which is used for reversing mis-predicted speculatively taken branches.

Every instruction destination is assigned a new physical register from free list. The sources are assigned the existing "Register Map tables" which is updated with the renamed destination. When register is moved from the free list to the active list, it is defined to be "busy" and "not busy" when its instruction completes and its result is stored in the register file. Thus busy-bit table determines if a result has been written into each of the physical registers.

For each operand, the "busy-bit table" is read and the bits are written into the queue with the instruction. The instruction waits if an operand is busy until it becomes "not busy." The integer and floating-point instructions are assigned registers from two separate free lists: integer and floating-point register files (each has 64 registers).

### 1.3.2.1 *Pipelining*

One of the main features of MIPS R10000 is the four-way superscalar pipelining where four instructions are handled in parallel. The MIPS R10000 has seven pipeline stages. The stage 1 fetches, stage 2 decodes. Stage 3 is used for issuing instructions and reading register operands and stages 4 through 6 executes instructions and finally stage 7 is used for storing the results.

Figure 1.10(b) gives the details of the pipeline stages of the MIPS R10000 microprocessor.

During stage 1 or the instruction fetch pipeline, four instructions are fetched in each cycle from the primary instruction cache (I-cache) independent of their alignment — except that the processor cannot fetch across a 16 word cache block boundary. These words are then aligned in a four-word Instruction register. They are pre-decoded in this stage to simplify the stage 2.

During stage 2, the instructions fetched in stage 1 are decoded and renamed. Integer and floating-point registers are renamed independently. Depending on the functional unit (integer, floating point, or load store) required, they are loaded into one of the three queues (integer, floating-point, and address). Register renaming occurs as part of the decoding process in which the logical registers are mapped into physical registers. Stage 2 can execute only one branch. The second branch instruction is not decoded until the next cycle.

In stage 3, instructions are issued and register operands are read. Decoded instructions are written into the queues. Stage 3 is also the start of each of the five execution pipelines.

In stages 4 through 6, instructions are executed in the various functional units. These units and their execution process are described below:

*Floating-point multiplier (three-stage pipeline).* This unit executes operations with single- or double-precision multiply with two-cycle latency and a one-cycle repeat rate. In the first two cycles, the multiplication is completed and the third cycle is used to pack and transfer the result.

*Floating-point divide and square-root units.* This unit executes operations with single- or double-precision division and square-root in parallel. Floating-point multiplier shares their issue and completion logic with these units.

*Floating-point adder (three-stage pipeline).* Add, subtract, compare, or convert operations are executed with this adder with a two-cycle latency and a one-cycle repeat rate. In the first two cycles, the operation is completed and in the third cycle it is used to pack and transfer the result.

*Integer ALU1 (one-stage pipeline).* This unit executes the operations such as integer add, subtract, branch and shift, and logic with a one-cycle latency and a one-cycle repeat rate. This ALU also verifies predictions made for branches that are conditional on integer register values. It is responsible for triggering rollback in case of branch misprediction.

*Integer ALU2 (one-stage pipeline).* Integer add, subtract, and logic operations are executed with a one-cycle latency and a one-cycle repeat rate. Integer multiplies and divide operations take more than one cycle.

*Load/store unit (address calculation and translation in the TLB).* By using either an integer or floating-point load or store instruction, a single memory address can be calculated every cycle. Address calculation and load operations can be calculated out of program order.

Finally, stage 7 is used to store the result in register file (RF).

More recent designs belonging to MIPS family have all been built on the R10000 core. The *R12000* uses a process to shrink the chip and run it at higher clock rates. The *R14000* has a speed of up to 600 MHz and adds support for DDR, SRAM in the cache. It has the computer bus speed of 200 MHz. The most recent version, the *R16000* has caches of 64 KB for both the instruction

# Uniprocessor Architecture Overview

and data cache and supports for up to 8 MB of level 2 cache, with the clock rates of 700 MHz.

## 1.4 Summary

The von Neumann single processor architecture was introduced followed by a review of the major enhancements made to this architecture. Details of two commercial architectures provided in this chapter illustrate the extent of pipeline and parallel structures utilized in contemporary architectures. The architectural features and performance characteristics of modern day processors change so fast that any book that attempts to capture them becomes obsolete by the time it is published. For details on latest processors and their characteristics, refer to corresponding manufacturer's manuals.

## Problems

The following problems may require a literature search. For details on recent architectures consult manufacturers' manuals and magazines such as:

IEEE Computer (monthly), Los Alamitos, CA: IEEE Computer Society.

Electronic Engineering Times (www.eetimes.com)

Electronic Design News (www.edn.com)

1.1. Trace the history of any computer system you have access to. Describe why there is a family rather than a single machine. Trace the evolution of each feature and the reason for change as the family evolved.

1.2. Answer Problem 1.1 with respect to the following popular families of computers:

   1. Motorola 680×0 series
   2. Intel 80×86 series
   3. International Business Machine Corporation's 370
   4. CRAY series from Cray Research Incorporated and Cray Computer Corporation
   5. Sun Microsystems workstations

1.3. There are several coprocessors (I/O, Numeric, and Memory Management) available now. Study their characteristics and determine how they support the corresponding CPUs.

1.4. Define "von Neumann bottleneck." To what degree the bottleneck has been minimized by the modern day structures?

1.5. What is an accepted definition of RISC architecture today?

1.6. Why did the following memory systems either proved to be unworkable or became obsolete:

 1. Magnetic bubble memories
 2. Magnetic drums
 3. Plated wire memories
 4. Magnetic core memories

1.7. Why did the following I/O devices become obsolete:

 1. Card reader
 2. Paper tape reader/punch
 3. Teletype

1.8. Look up the definitions of the following: parallelism, pipelining, and overlapped processing.

1.9. Look up the definitions of the following terms: multiprogramming, multiprocessing, multitasking, batch processing, time-sharing.

1.10. For each subsystem of a single-processor computer system:

 1. Identify the forms of parallelism possible
 2. List the hardware, software, and firmware support needed for each form, and
 3. Identify commercially available systems that support each form.

1.11. Show the schematic of a $(16K \times 16)$ memory system, built out of $(1K \times 8)$ memory modules using (a) high-order interleaving and (b) low-order interleaving.

1.12. Three memory chips are available with the following characteristics:

|  | Chip 1 | Chip 2 | Chip 3 |
|---|---|---|---|
| Relative cost | 1 | 2 | 4 |
| Read/write speed (clock pulses) | 4 | 2 | 1 |

Assume that due to their low cost, a wider word fetch scheme that fetches 4 and 2 words per cycle can be used with chips 1 and 2, respectively. Chip 3 implementation is cost-effective only for single word fetch. Also, assume that in 25% of the fetch cycles, half of the words fetched in wider word fetch schemes are useless. Estimate the relative bandwidths of the three systems.

1.13. Assume that the cost of the CPU is 25% of the total cost of a computer system. It is possible to increase the speed of the CPU by a factor of 10 by increasing the cost also by 10 times. The CPU typically waits for I/O about 30% of the time. From a cost/performance viewpoint, is increasing the speed tenfold desirable?

## References

Burks, A. W., Goldstine, H. H., and von Neumann, J., Preliminary Discussion of the Logical Design of an Electrical Computing Instrument, *U.S. Army Ordinance Department Report*, 1946.

*Intel Itanium Architecture Software Developer's Manual*, December 2001, http://devresource.hp.com/drc/STK/docs/refs/24531703s.pdf.

*Intel Itanium Architecture Software Developer's Manual, Vol.1: Application Architecture*, October 2002, http://developer.intel.com/design/itanium/manuals/245317.pdf.

*Intel Itanium Processor, Development Tools*, http://www.intel.com/design/itanium/devtools/.

*Intel Itanium 2 Processor, Hardware Developer's Manual*, July 2002, http://www.intel.com/design/itanium2/manuals/25110901.pdf.

*MIPS R10000 Microprocessor User's Manual*, Version 2.0, January 29, 1997, http://techpubs.sgi.com/library/manuals/2000/007-2490-001/pdf/007-2490-001.pdf.

*MIPS R10000 Microprocessor User's Manual*, Version 2.0, October 10, 1996, http://cch.loria.fr/documentation/docSGI/R10K/t5.ver.2.0.book.pdf.

MIPS Technologies, http://www.mips.com.

Shiva, S. G., *Computer Design and Architecture*, 3rd ed. New York: Marcel Dekker, 2000.

# 2

## Models and Terminology

Chapter 1 has traced the evolution of architectures as enhancements to von Neumann's uniprocessor model. These enhancements employed several parallel and overlapped processing concepts. For instance, the DMA (direct memory access) controller performing the input/output (I/O) concurrent with the processing activities of the CPU (central processing unit), and multiple functional units in the ALU (arithmetic logic unit) subsystem operating concurrently, are examples of parallel-processing structures. The pipelined implementation of ALUs and control units is an example of overlapped processing structure. Although these structures enhance the performance of the computer system, it is still viewed as a serial system in the sense that the parallelism in the application is not completely exploited. That is, the hardware structures are designed to execute the application essentially in a serial mode while attaining the highest possible performance.

As mentioned in Chapter 1, there are applications that make the system throughput inadequate even with all the enhancements. The processing speed required by these applications cannot be achieved by the fastest CPU available. As such, newer architectures that can provide higher performance and circumvent the limits of technology have evolved. These parallel-processing architectures are designed to better exploit the parallelism in the application. This chapter introduces parallel-processing concepts, terminology and architectures at a very gross level of detail. Subsequent chapters of this book provide further details on each of these architectures.

Sections 2.1 through 2.3 of this chapter expand the terminology and basic models of advanced architectures introduced in Chapter 1. With the evolution of architectures, several architecture classification schemes (*taxonomies*) have also evolved. The popular taxonomy due to Flynn (1966) is utilized in Section 2.4 to provide the framework for the description of popular architectures. Sections 2.5 and 2.6 deal with performance and cost issues.

## 2.1 Effect of Application on the Architecture

Suppose that the application allows the development of processing algorithms with a degree of parallelism $A$. The *degree of parallelism* is simply the number of computations that can be executed concurrently. Further, if the language used to code the algorithm allows the representation of algorithms with a degree of parallelism $L$, the compilers produce an object code that retains a degree of parallelism $C$ and the hardware structure of the machine has a degree of parallelism $H$, then, for the processing to be most efficient, the following relation must be satisfied:

$$H \geq C \geq L \geq A \qquad (2.1)$$

Here, the objective is to minimize the computation time of the application at hand. Hence, the processing structure that offers the least computation time is the most efficient one.

For the architecture to be most efficient, the development of the application algorithms, programming languages, the compiler, the operating system, and the hardware structures must proceed together. This mode of development is only possible for very few special purpose applications. In the development of general-purpose architectures however, the application characteristics cannot be easily taken into account. But, the development of other components should proceed concurrently, as far as possible.

Development of algorithms with a high degree of parallelism is application dependent and basically a human endeavor. A great deal of research has been devoted to developing languages that contain parallel-processing constructs, thereby enabling the coding of parallel algorithms. Compilers for these parallel-processing languages retain the parallelism expressed in the source code during the compilation process, thus producing parallel object code. Also, compilers that extract parallelism from a serial program (thus producing a parallel object code) have been developed. Progress in hardware technology has yielded a large number of hardware structures that can be used in executing parallel code. Subsequent chapters of this book provide details of such parallel architectures along with pertinent programming language and compiler concepts.

## 2.2 Application Characteristics

In order to obtain the best possible performance from a parallel-processing system, the application at hand should be partitioned into a set of *tasks* that can be executed concurrently. Each task accomplishes a part of the overall computation and typically requires some sort of communication (i.e., to transmit results, to request or provide data items that may be common to both tasks, etc.) with other tasks during execution.

In evaluating the performance of such implementations, four application characteristics are generally considered important:

1. Granularity
2. Degree of parallelism

3. Level of parallelism
4. Data dependency

The *granularity* of an application module (task) is a function of $R$, the run time (execution time) of the task and $C$, the communication time of the task with other tasks. If $R$ is large compared to $C$, the task granularity is said to be *coarse* (large). That is, a coarse-grain task has the least communication overhead and tasks in a coarse-grained application spend most amount of time in performing useful computations and least amount in communicating with each other. For *fine*-grained tasks, $C$ dominates $R$. That is, the task performs a small amount of computation before requiring communication with other tasks. A medium-grain task is the compromise between the above two extremes.

The *degree of parallelism* (defined earlier) exhibited by the application is a measure of the number of threads of computation that can be carried out simultaneously.

The *level of parallelism* in a way dictates the granularity. The following levels (with decreasing granularity) are typically considered:

1. Procedure level
2. Task level
3. Instruction level
4. Operation level
5. Microcode level

Here, the assumption is that the application is composed of several procedures that can be executed in parallel, thus resulting in procedure level parallelism. Each application is considered to have multiple tasks, which can be executed in parallel. Instruction level parallelism allows the concurrent execution of instructions in a given task. An instruction is composed of several operations.

---

**Example 2.1**
The instruction:
$$F = A * B + C * D \qquad (2.2)$$
consists of two multiplications, one addition, and one assignment operation. Each operation is brought about by a set of microoperations (microcode).

---

The order in which the procedure and task levels occur in the above list is some times reversed in the literature, since there is no well accepted "inclusion" between procedures and tasks. Further, some do not distinguish between instruction and operation levels. In general, the microcode level of parallelism is not of interest to an application designer. But, it is of interest to the hardware and microcode designer.

Consider again the computation in (2.2). Here, the two multiplications can be performed in parallel, if two processors are available. But, the addition cannot be performed until the multiplications are completed and the assignment operation cannot be performed until the addition is complete. Thus, even if the architecture has four processors available, the maximum number of processors that can be utilized concurrently for this application is two (i.e., the degree of parallelism is 2), because of the *precedence constraints* between the operations. These constraints are obeyed automatically in a single processor system since only one operation is executed at a time. When the application is split into several tasks to be implemented on a multiple processor system, a task requiring a data item will have to wait until another task producing it has indeed produced that item. Further, if several tasks are updating a (shared) variable, the update sequence should be such that the result is same as that would be obtained when the application is executed by a single processor system. Thus, *Data dependencies* are the result of precedence constraints between operations (or tasks, or procedures) imposed by the application. Data dependencies also affect the granularity and the degree of parallelism of the application.

## 2.3 Processing Paradigms

The main objective of multiple processor architectures is to minimize the computation time (i.e., to maximize the throughput of the system) of an application. Ideally, a computer system with $N$ processors should provide a throughput of $N$ times that of a single processor system. This requires that the degree of parallelism of the application is almost $N$.

In practice, the communication between tasks (due to data dependencies), allocation of tasks to processors and, controlling the execution of multiple tasks generate considerable overhead. Hence, the $N$-fold speed up mentioned above is not possible, in general.

There are applications however, which can be partitioned such that there is no communication between tasks. These are called *trivially parallel* applications. If the system structure minimizes the task creation and execution overhead, an $N$-fold throughput can be achieved for these applications.

One obvious way of reducing the overhead is to partition the application into tasks with as large a granularity as possible. Although this minimizes the communication and task management overhead, it reduces the degree of parallelism since there are now fewer tasks that can be executed concurrently. To maximize the degree of parallelism, the application must be partitioned into as large a number of fine-grain tasks as possible. But, this increases the execution overhead. Thus, a compromise is needed between the granularity and the degree of parallelism, to maximize the throughput of the system.

# Models and Terminology

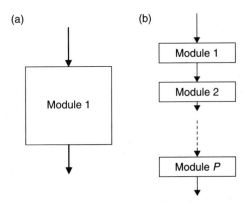

**FIGURE 2.1**
Completely serial application: (a) single task and (b) multiple tasks

The above discussion assumed that throughput enhancement is the objective for building parallel architectures. Another common objective is to maximize the hardware utilization. In this context, the efficiency of the system is measured by the percentage of time during which all the processors are busy.

There are a variety of ways in which an application can be modeled, of which the following are the most general ones:

1. Completely serial
2. Serial–parallel–serial without data dependencies
3. Serial–parallel–serial with data dependencies

Figure 2.1 shows the model of a *completely serial* application. Here the degree of parallelism is 1. In Figure 2.1(a), the application is implemented as a single task utilizing a single processor. The total computation performed (i.e., the total number of instructions or operations) by the task is $W$. In Figure 2.1(b), the task is implemented as $P$ sequential subtasks each performing a part ($W_i$) of the work $W$. Thus the total work is the sum of the work performed by the subtasks:

$$W = \sum_{i=1}^{P} W_i \qquad (2.3)$$

The latter model is useful when the task needs to be spread over multiple processors due to resource limitations of the individual processor in the system.

Figure 2.2 shows the *serial–parallel–serial* (without data dependencies) model. Here, the first subtask initiates the computational task and spawns $P$-2 subtasks. The subtasks are executed independently and the results are combined by the last subtask. The total work $W$ is now performed by $P$ subtasks

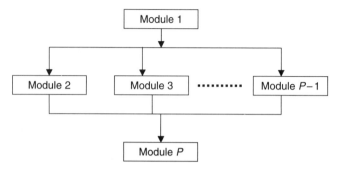

**FIGURE 2.2**
Serial–parallel–serial application without data dependencies

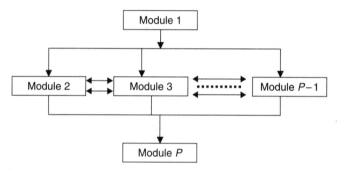

**FIGURE 2.3**
Serial–parallel–serial application with data dependencies

with $P$-2 of them executing concurrently. This is the case of the so called *easy* (or *trivial*) *parallelism* since the $P$-2 subtasks can execute to completion once they are spawned. This model can also be viewed as a *supervisor/worker* model, where task 1, the supervisor spawns the $P$-2 worker tasks. The results produced by these workers are collected by the supervisor (as task $P$). The overhead introduced by this model is due to: spawning and subtask creation, subtask allocation to processors, and orderly termination of the $P$-2 subtasks. This model is ideal for applications that can be partitioned into subtasks with no data dependency between them. That is, the data structure is also partitioned such that each subtask has its own independent data to operate upon. There is no sharing of data between the concurrent subtasks.

Figure 2.3 shows the *serial–parallel–serial with data dependencies* model. Here subtasks communicate with each other passing messages to resolve data dependencies. Typically, the processor executing the subtask can continue its processing for a while, once a request for a data item is sent out. If the requested data did not arrive, the processor waits for it instead of switching to another task, in order to minimize task switching overhead. This is also called the *communication-bound* model.

It is important to note that an application might exhibit all the above modes of computation in practice. That is, an application can be partitioned into several procedures where the set of procedures might fit one of the above models. Each procedure in turn, may be partitioned into tasks, which again would fit one of the three models, and so on. Such partitioning is obviously influenced by the architecture of the underlying hardware.

Several other parallel-processing paradigms have evolved along with appropriate architectures, over the years. The most common ones are described next starting with Flynn's taxonomy.

## 2.4 Flynn's Taxonomy

Before providing the details of Flynn's taxonomy (Flynn, 1966), it is interesting to examine the need for taxonomies or classification schemes. Skillicorn (1988) provides three reasons for the classification of computer architectures:

1. The classification can answer questions such as, what kind of parallelism is employed, which architecture has the best prospect for the future, and what has already been achieved by the architecture in terms of particular characteristics one is interested in.
2. The classification reveals possible configurations that might not have otherwise occurred to the system architect. That is, it allows a formal examination of possible design alternatives, once the existing architectures have been classified.
3. The classification allows the building of useful models of performance, thereby revealing the potential of a particular architecture for improvement in performance.

Taxonomy is usually based on one or more pertinent characteristics of the systems that are being classified. For example, Milutinovic (1989) combines the taxonomies by Treleaven et al. (1982) and Myers (1982) to arrive at the following taxonomy, based on what "drives" the computational flow of the architecture:

1. *Control-driven (control-flow) architectures*:
    (a) Reduced instruction set computers (RISC)
    (b) Complex instruction set computers (CISC)
    (c) High-level language (HLL) architectures
2. *Data-driven (data-flow) architectures*
3. *Demand-driven (reduction) architectures*

In *control-driven* or *control-flow architectures*, the instruction sequence (i.e., the program) guides the processing activity. That is, the flow of computation is determined by the instruction sequence and data are gathered as and when an instruction needs them. All architectures introduced in Chapter 1 and all commercial architectures to date belong to this category. Chapters 3 through 6 concentrate on this class of architectures.

The advent of VLSI technology provided the capability to fabricate a complete processor on an IC chip. An analysis of such ICs indicates that implementation of the control unit of the processor consumes 60–70% of the chip area, thereby limiting the number of processor functions that can be implemented on the chip. Since the control unit complexity is proportional to the number of instructions in the instruction set, an obvious way to reduce the control unit complexity is to reduce the number of instructions in the instruction set. Simplification of the control unit to enable building a complete processor on an IC chip was the prime motivation behind the *RISC* designs first initiated at IBM in 1975 and in 1980 at the University of California, Berkeley. Obviously, the most frequently used instructions were selected and the control unit was optimized to provide the fastest possible execution.

The progress in IC technology also contributed to the design of complex instruction set computers (*CISCs*). Several HLL constructs became part of the machine instruction set. But, designing compilers that utilize these high-level constructs in the production of object code from the HLL program became a complex task. Since most of the compilers utilized only 60–70% of the constructs found in the instruction set, implementation of elaborate instruction sets was deemed unnecessary, supporting the idea of simpler control units.

Consider for example, the instruction set of Digital Equipment Corporation (DEC) VAX-11, a CISC architecture. It consists of 304 instructions with 16 addressing modes utilizing 16 registers. It supports a considerable number of data types (6 types of integers, 4 types of floating-points, packed decimal and character strings, variable length bit fields, etc.) and variable length instructions (2 to 25 bytes) with up to six operand specifiers.

An RISC is expected to have much less than 304 instructions and much simpler instruction environment. There is no consensus about the number of instructions in an RISC instruction set. The Berkeley RISC-I had 31, the Stanford University MIPS had over 60, and the IBM 801 had over 100. The simplification is achieved also by reducing the number of addressing modes and the number of instruction formats.

The RISC designs with more than 100 instructions are now on the market and the characteristics that distinguish an RISC from a CISC are getting blurred. Nevertheless the following items provide a framework to characterize RISC architectures:

1. A relatively low number of instructions (around 100).
2. A small number of addressing modes (around three).
3. A small number of instructions format (mostly fixed length).

4. Fast execution of all instructions by utilizing a large number of parallel data paths and overlapping the fetch and execute phases of successive instructions (i.e., pipelined execution).
5. Minimized memory access, by utilizing a large number of registers and mostly register-to-register instructions. Memory access is performed by only load and store instructions.
6. Support for most frequently used operations in the application inherently in the machine design, by a judicious choice of instructions and optimizing compilers.

The aim of RISCs is to implement a powerful processor while maintaining a small instruction set. These architectures depend on the compilers for the reduction of software semantic gap. That is, a HLL program has to be translated into the machine primitives represented by the small instruction set. HLL architectures on the other hand, tend to reduce the semantic gap by making the primitives available in the instruction set identical to those in the HLL. That is, the hardware interprets the HLL constructs directly.

When HLLs are used for programming computers, the programs must be converted into the object code (or machine language program) before the program execution begins. In conventional architectures, the translation from the HLL source to machine language is performed by software means (i.e., compilers) and the execution is performed by the hardware. Introduction of block-structured languages helped the speed up of the compilation process. Machine hardware supported the translation process of such languages by stack organizations. As the hardware costs came down with the improvements in hardware technology, more complex instructions (e.g., block move, loop) were added to the assembly language instruction sets to provide a more direct translation capability. In HLL architecture the hardware is specifically designed for translating and subsequently executing the source HLL program. A well-known architecture of this type is the SYMBOL machine developed at Iowa State University in 1971, to directly execute a HLL called Symbol programming language.

The advantage of HLL architectures is a very high translate-load speed. It has not been proven that they offer execution speeds higher than conventional architectures. A further disadvantage is that only one source language for which the machine is designed can be used in programming the machine. For architecture to be commercially successful, it has to be general purpose enough to suit a large number of applications, thus increasing the number of units sold and reducing the cost per unit. As such, HLL architectures have never been successful commercially.

*Data-driven* or *dataflow architecture* is controlled by the readiness of data. That is, an instruction is executed (i.e., the control activity is invoked) when the data required by that instruction are ready to be operated upon. In *reduction* or *demand-driven architectures* an instruction is enabled for execution when its results are required as operands for another instruction that has already been enabled for execution. These experimental architectures have not seen commercial success. Data-driven architectures are introduced in Chapter 7.

Flynn's taxonomy is based on the degree of parallelism exhibited by architecture in its data and control flow mechanisms. He divides computer architectures into four main classes shown in Figure 2.4, based on the number of instruction and data streams:

1. Single instruction stream, single data stream (SISD) machines, which are single processor systems (uniprocessor).

2. Single instruction stream, multiple data stream (SIMD) architectures, which are systems with multiple arithmetic-logic processors and a single control processor (CP). Each arithmetic-logic processor processes a data stream of its own, as directed by the single CP. This classification includes *array processors*.

3. Multiple instruction stream, single data stream (MISD) machines, in which the single data stream is simultaneously acted upon by a multiple instruction stream. This classification is considered an aberration, since it is not practical to implement such an architecture. (Note that in pipelined processors multiple instruction streams exist. If the definition of the single data stream is stretched to include the conglomeration of all the data elements in pipeline stages, then pipelined processors fit this classification.)

4. Multiple instruction stream, multiple data stream (MIMD) machines, which contain multiple processors, each executing its own instruction stream to process the data stream allocated to it (i.e., a *multiprocessor system*). A computer system with a central processor and an I/O processor working in parallel is the simplest example of MIMD architecture.

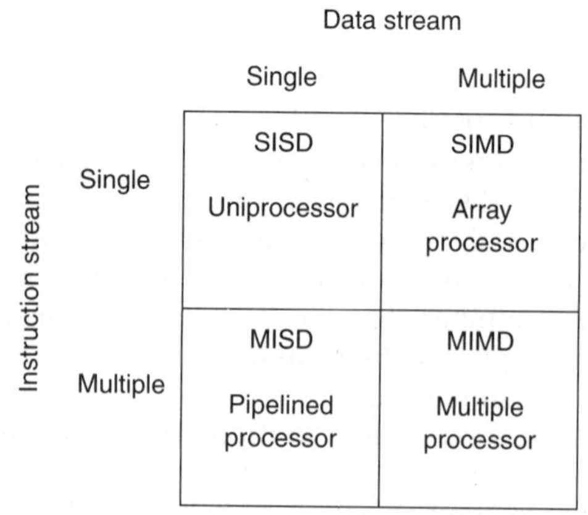

|  | Data stream | |
|---|---|---|
|  | Single | Multiple |
| Instruction stream — Single | SISD<br>Uniprocessor | SIMD<br>Array processor |
| Instruction stream — Multiple | MISD<br>Pipelined processor | MIMD<br>Multiple processor |

**FIGURE 2.4**
Flynn's architecture classification

## Models and Terminology

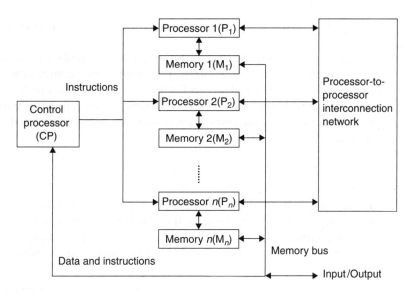

**FIGURE 2.5**
SIMD structure

Chapter 1 provided the details of SISD architectures. Models for SIMD and MIMD classifications are provided below and further details on these architectures are provided in subsequent chapters.

### 2.4.1 Single Instruction Stream, Multiple Data Stream

Figure 2.5 shows the structure of a typical SIMD machine. There are $n$ arithmetic-logic processors ($P_1$–$P_n$), each with its own memory block ($M_1$–$M_n$). The individual memory blocks combined constitute the system memory. A bus is used to transfer instructions and data to the CP from the memory blocks. The CP decodes instructions and sends control signals to processors $P_1$–$P_n$.

The CP, in practice, is a full fledged uniprocessor. It retrieves instructions from memory, sends arithmetic-logic instructions to processors, and executes control (branch, stop, etc.) instructions itself. Processors $P_1$–$P_n$ executes the same instruction, each on its own data stream. Based on arithmetic-logic conditions, some of the processors may be deactivated during certain instructions. Such activation and deactivation of processors is handled either by the CP or by the logic local to each arithmetic processor.

Some computations on SIMD require that the data be exchanged between arithmetic processors. The processor interconnection network enables such data exchange. Chapters 5 and 6 provide further details on such networks.

The most important characteristic of SIMDs is that the arithmetic processors are synchronized at the instruction level. That is, they execute programs in a "lock step" mode, where each processor has its own data stream. They are also called *data parallel architectures*.

The SIMDs are special purpose machines, albeit with a wide variety of suitable applications. They are called *array* processors since computations involving arrays of data are natural targets for this class of architecture. (Note that the term "array processor" has been used to describe architectures with an array of processors even though these processors are not synchronized at the instruction level.)

---

**Example 2.2**
Consider the computation of the column sum of a matrix. Each column of the matrix can be assigned to one of the $n$ arithmetic processors of an SIMD. The column sums of an $N \times N$ matrix can then be computed in $N$ steps (where $N \leq n$), rather than in $N^2$ steps required on an SISD machine.

---

As illustrated by the above example, SIMD systems can provide a high throughput, as long as the processing algorithm exhibits a high degree of parallelism at the instruction level.

Illiac-IV and Goodyear Aerospace's STARAN are SIMDs of historical interest. Thinking Machine Corporation's Connection Machine (models 1 and 2) and Maspar Corporation's MP series are examples of recent commercial SIMDs that have become obsolete. Modern microprocessors such as Intel's Itanium offer SIMD mode of operation. Chapter 5 provides further details on SIMD machines.

### 2.4.2 Multiple Instruction Stream, Multiple Data Stream

Figure 2.6 shows an MIMD structure consisting of $p$ memory blocks, $n$ processing elements, and $m$ I/O channels.

The processor-to-memory interconnection network enables the connection of a processor to any of the memory blocks. Since all the memory blocks are accessible by all the processors, this is known as the *shared memory* MIMD architecture. Ideally, $p$ should be greater than or equal to $n$ and the interconnection network should allow $p$ simultaneous connections, in order that all the processors are kept busy. In addition, it should be possible to change these connections dynamically, as the computation proceeds.

The processor-to-I/O interconnection network enables the connection of any I/O channel to any of the processors.

The processor-to-processor interconnection network is an interrupt network to facilitate data exchange between the processors. A processor generating data places them in the memory and interrupts the processor that requires that data, through this network.

## Models and Terminology

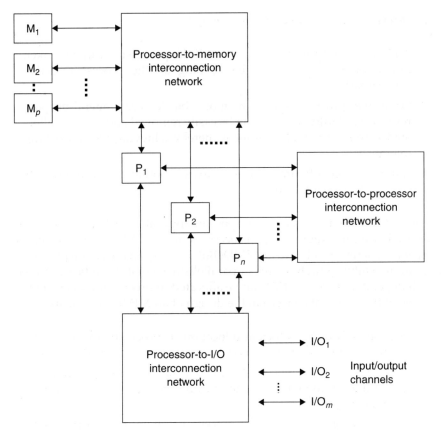

**FIGURE 2.6**
MIMD structure

**Example 2.3**
Consider the computation of the column sum of an $N \times N$ matrix. This can be partitioned into $N$ tasks. The program to accumulate the elements in a column and the corresponding data form a task. Each task is assigned to a processor in the MIMD. If the MIMD contains $N$ processors, all the $N$ tasks can be performed simultaneously, thus completing the computation in $N$ steps, rather than the $N^2$ steps needed by an SISD. If the number of processors $p$ is less than $N$, $p$ tasks are performed simultaneously in one cycle, requiring $N/p$ such cycles to complete the computation. Tasks are assigned to processors as they become free after completing the task currently assigned to them. There is a considerable amount of coordination overhead required to execute tasks in this manner.

The MIMDs offer the following advantages:

1. A high throughput can be achieved if the processing can be broken into parallel streams, thereby keeping all the processors active concurrently.
2. Since the processors and memory blocks are general-purpose resources, a faulty resource can be easily removed and its work allocated to another available resource, thereby achieving a degree of fault tolerance.
3. A dynamic reconfiguration of resources is possible to accommodate varying processing loads.

The MIMDs are more general-purpose in application than are SIMDs, but they are harder to program. The processors in an MIMD are not synchronized at instruction-level as in an SIMD. But, it is required that the processing algorithm exhibits a high degree of parallelism, so that it can be partitioned into independent subtasks that can be allocated to processors concurrently. Some of the issues of concern in the design of an MIMD system are:

1. *Processor scheduling*: efficient allocation of processor to processing needs in a dynamic fashion as the computation progresses.
2. *Processor synchronization*: prevention of processors trying to change a unit of data simultaneously and obeying the precedence constraints in data manipulation.
3. *Interconnection network design*: processor-to-memory and processor-to-processor interconnection networks are probably the most expensive elements of the system. Because of cost considerations, almost all MIMDs to date have not used a processor-to-I/O network, rather they dedicate I/O devices to processors.
4. *Overhead*: ideally an $n$ processor system should provide $n$ times the throughput of a uniprocessor. This is not true in practice because of the overhead processing required to coordinate the activities between the various processors and to resolve the contention for resources (such as memory and interconnection network).
5. *Partitioning*: identifying parallelism in processing algorithms to invoke concurrent processing streams is not a trivial problem.

Several experimental MIMD systems have been built. Carnegie Mellon University's C.mmp and Cm*, New York University's Ultracomputer, University of Illinois' Cedar are some early examples. Some popular commercial MIMD systems are: BBN Butterfly, Alliant FX series, Thinking Machine Corporation's CM-5, and Intel corporation's iPSc series that have become obsolete. Many new MIMD systems are being introduced every year. Chapter 6 provides further details.

It is important to note that Flynn's architecture classification is not unique in the sense that a computer system may not clearly belong to one of the three classes. For example, the CRAY series of supercomputers can be classified under all three classes, based on their operating mode at a given time.

## 2.5 Computer Networks

There are three aspects in which conventional SIMD and MIMD approaches deemed to fall short:

1. They are not highly scalable and hence cannot be easily expanded in small increments.
2. They employ only one type of general-purpose processor and hence are not suitable for environments with an array of specialized applications.
3. They employ fixed and *tightly coupled* interconnection topology, thereby restricting the users when the applications dictate a different more efficient topology.

Recent SIMD and MIMD systems have addressed the first two shortfalls by using heterogeneous processing nodes and offering scalability to fairly large number of nodes. They have also merged the SIMD and MIMD concepts. In fact, the evolution of the Thinking Machines CM series illustrates this. The earlier machines in the series were SIMDs while the CM-5 operates in both the modes.

*Computer networks* are the most common multiprocessor architectures today. Figure 2.7 shows the structure of a computer network. It is essentially an MIMD system, except that the nodes are *loosely coupled*, by the communication network. Each node (Host) is an independent computer system. The user can access the resources at the other nodes through the network. The important concept is that the user executes his application at the node he is connected to, as far as possible. He submits his job to other nodes when resources to execute the job are not available at his node. The *Internet* is the best example of the world wide network.

A *distributed processing system* utilizes the hardware structure of the network shown in Figure 2.7. It is more general in its operation in the sense that the nodes operate in what is called a "cooperative autonomy." That is, the nodes cooperate to complete the application at hand and all the nodes are equal in capability (i.e., there is no master/slave relation among the nodes). In a distributed system, the hardware, the data, and the control (operating system) are all distributed.

The next step in this progression of building powerful computing environments is *Grid computing*. A grid is a network of computers that act as a single "virtual" computer system. Utilizing specialized scheduling software,

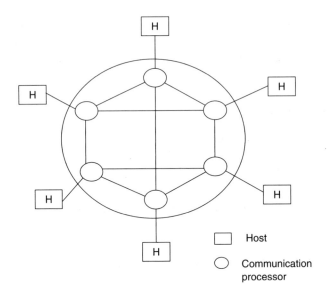

**FIGURE 2.7**
Computer network

grids identify resources and allocate them to tasks for processing on the fly. Resource requests are processed wherever it is most convenient, or wherever a particular function resides, and there is no centralized control. Grid computing exploits the underlying technologies of distributed computing, job-scheduling, and workload management, which have been around more than a decade. The recent trends in hardware (commodity servers, blade servers, storage networks, high-speed networks, etc.) and software (Linux, Web services, open source technologies, etc.) have contributed to make grid computing practical. Hewlett–Packard's Adaptive Enterprise initiative, IBM's On Demand Computing effort, Sun Microsystems's Network One framework are examples of commercial grid computing products. Chapter 7 provides further details on Grid computing.

## 2.6 Performance Evaluation

Several measures of performance have been used in the evaluation of computer systems. The most common ones are: million instructions per second (MIPS), million operations per second (MOPS), million floating-point operations per second (MFLOPS or megaflops), billion floating-point operations per second (GFLOPS or gigaflops) , and million logical inferences per second (MLIPS). Machines capable of trillion floating-point operations per second (teraflops) are now available.

The measure used depends on the type of operations one is interested in, for the particular application for which the machine is being evaluated. As such, any of these measures have to be based on the mix of operations representative of their occurrence in the application.

---

**Example 2.4**
The instruction mix in an application and instruction execution speeds of a hypothetical machine are as follows:

| Instruction | Speed (cycles) | Occurence (%) |
|---|---|---|
| ADD | 8 | 30 |
| SHIFT | 4 | 20 |
| LOAD | 12 | 30 |
| STORE | 12 | 20 |

Thus, the average instruction speed is ($8 \times 0.3 + 4 \times 0.2 + 12 \times 0.3 + 12 \times 0.2$) = 9.2 cycles. If the clock frequency is 1 MHz (i.e., 1 million cycles per second), the machine performs 1/9.2 MIPS. This rating is more representative of the machine performance than the maximum rating (1/4 MIPS) computed by using the speed of execution (4 cycles) of the fastest instruction (SHIFT).

---

Thus, the performance rating could be either the *peak rate* (i.e., the MIPS rating the CPU cannot exceed) or the more realistic *average or sustained rate*. In addition, a comparative rating that compares the average rate of the machine to that of other well-known machines (e.g., IBM MIPS, VAX MIPS, etc.) is also used.

In addition to the performance, other factors considered in evaluating architectures are: *generality* (how wide is the range of applications suited for this architecture), *ease of use*, and *expandability* or *scalability*. One feature that is receiving considerable attention now is the *openness* of the architecture. The architecture is said to be *open* if the designers publish the architecture details such that others can easily integrate standard hardware and software systems to it. The other guiding factor in the selection of architecture is the *cost*. Some of the factors contributing to the cost parameter are explored in Section 2.7.

Several analytical techniques are used in estimating the performance. All these techniques are approximations and as the complexity of the system increases, most of these techniques become unwieldy. A practical method for estimating the performance in such cases is by using benchmarks.

### 2.6.1 Benchmarks

Benchmarks are standardized batteries of programs run on a machine to estimate its performance. The results of running a benchmark on a given machine can then be compared with those on a known or standard machine, using criteria such as CPU and memory utilization, throughput and device utilization, etc.

Benchmarks are useful in evaluating hardware as well as software and single processor as well as multiprocessor systems. They are also useful in comparing the performance of a system before and after certain changes are made.

As a HLL host, a computer architecture should execute efficiently those features of a programming language that are most frequently used in actual programs. This ability is often measured by benchmarks. Benchmarks are considered to be representative of classes of applications envisioned for the architecture. Some common benchmarks are:

1. *Real world/application benchmark.* These use system- or user-level software code drawn from real algorithms or full applications, commonly used in system-level benchmarking. These usually have large code and data storage requirements.

2. *Derived benchmarks.* These are also called as "algorithm-based benchmarks." These extract the key algorithms and generate realistic datasets from real world applications. These are used for debugging, internal engineering, and for competitive analysis.

3. *Single processor benchmark.* These are low-level benchmarks used to measure performance parameters that characterize the basic architecture of the computer. These hardware/compiler parameters predict the timing and performance of the more complex kernels and applications. These are used to measure the theoretical parameters that describe the overhead or potential bottleneck, or the properties of some item of hardware.

4. *Kernel benchmarks.* These are code fragments extracted from real programs in which the code fragment is responsible for most of the execution time. These have the advantage of small code size and long execution time. Examples are *Linpack* and Lawrence Livermore loops. The Linpack measures the MFLOPS rating of the machine in solving a system of equations in a FORTRAN environment. The *Lawrence Livermore* loops measure the MFLOPS rating in executing 24 common FORTRAN loops operating on datasets with 1001 or fewer elements.

5. *Local benchmarks.* These are programs that are site-specific. That is, they include in-house applications that are not widely available. Since the user is most interested in the performance of the machine for his or her applications, local benchmarks are the best means of evaluation.

6. *Partial benchmarks.* These are partial traces of programs. It is in general difficult to reproduce these benchmarks when the portion of benchmarks that was traced is unknown.

7. *UNIX utility and application benchmarks.* These are programs that are widely employed by the UNIX user community. The SPEC (System Performance Evaluation Cooperative Effort) Benchmark suite belongs to this category and consists of ten scenarios taken from a variety of science and engineering applications. This suite developed by a consortium of computer vendors is

for the evaluation of workstation performance. The performance rating is provided in *SPECmarks*.

8. *Synthetic benchmarks*. These are small programs constructed specially for benchmarking purposes. They do not perform any useful computation, but statistically approximate the average characteristics of real programs. Examples are Dhrystone and *Whetstone* benchmarks.

The Whetstone benchmark, which in its original form was developed in ALGOL 60. Whetstone reflects mostly numerical computing, using a substantial amount of floating-point arithmetic. It is now chiefly used in a FORTRAN version. Its main characteristics are as follows:

1. A high degree of floating-point data and operations, since the benchmark is meant to represent numeric programs.
2. A high percentage of execution time is spent in mathematical library functions.
3. Use of very few local variables, since the issue of local versus global variables was hardly being discussed when these benchmarks were developed.
4. Instead of local variables, a large number of global variables are used. Therefore, a compiler in which the most heavily used global variables are used as register variables (as in 3) will boost the Whetstone performance.
5. Since the benchmark consists of nine small loops, Whetstone has an extremely high code locality. Thus, a near 100% hit rate can be expected even for fairly small instruction caches.

The distribution of the different statement types in this benchmark was determined in 1970. As such, the benchmark cannot be expected to reflect the features of more modern programming languages (e.g., Record and pointer data types). Also, recent publications on the interaction between programming languages and architecture have examined more subtle aspects of program behavior (e.g., the locality of data references-local versus global) that were not explicitly considered in earlier studies.

In early efforts dealing with the performance of different computer architectures, performance was usually measured using some collection of programs that happened to be available to the user. However, following the pioneering work of Knuth in early 1970s an increasing number of publications have been providing statistical data about the actual usage of programming language features. The *Dhrystone benchmark* program set is based on these recent statistics, particularly in systems programming. Its main features are as follows:

6. It contains a measurable quantity of floating-point operations.
7. A considerable percentage of execution time is spent in string functions. In case of C compilers this number goes up to 40%.

8. Unlike Whetstone, Dhrystone contains hardly any loops within the main measurement loop. Therefore, for processors with small instruction caches, almost all the memory accesses are cache misses. But as the cache becomes larger, all the accesses become cache hits.
9. Only a small amount of global data is manipulated and the data size cannot be scaled.

9. *Parallel benchmarks.* These are for evaluating parallel computer architectures. The 1985 workshop at the National Institute of Standards (NIST) recommended the following suite for parallel computers: Linpack, Whetstone, Dhrystone, Livermore loops, Fermi National Accelerator Laboratory codes used in equipment procurement, NASA/Ames benchmark of 12 Fortran subroutines, John Rice's numerical problem set and Raul Mendez's benchmarks for Japanese machines.

10. *Stanford small programs.* Concurrent with the development of the first RISC systems John Hennessy and Peter Nye at Stanford's Computer systems laboratory collected a set of small C programs. These programs became popular because they were the basis for the first comparisons of RISC and CISC processors. They have now been collected into one C program containing eight integer programs (Permutations, Towers of Hanoi, Eight queens, Integer matrix multiplication, Puzzle, Quicksort, Bubble sort, and Tree sort) and two floating-point programs (matrix multiplication and fast Fourier transform).

11. *PERFECT.* The PERFormance Evaluation for Cost-effective Transformations benchmark suite consists of 13 Fortran subroutines spanning four application areas (signal processing, engineering design, physical and chemical modeling, and fluid dynamics). This suite consists of complete applications (with the I/O portions removed) and hence constitutes significant measures of performance.

12. *SLALOM.* The Scalable, language-independent, Ames Laboratory, 1-min measurement is designed to measure the parallel computer performance as a function of problem size. The benchmark always runs in 1 min. The speed of the system under test is determined by the amount of computation performed in 1 min.

There are many other benchmark suites in use and more are being developed. It is important to note that the benchmarks provide only a broad performance guideline. It is the responsibility of the user to select the benchmark that comes close to his application and further evaluate the machine based on scenarios expected in the application for which the machine is being evaluated.

## 2.7 Cost Factor

The unit cost of the machine is usually expressed as dollars per MIPS (or MFLOPS). It is important to note that the cost comparison should be performed on architectures of approximately the same performance level. For example, if the application at hand requires a performance level of $N$ MIPS, it is usually an overkill to select an architecture that delivers $M$ MIPS where $M$ is far greater than $N$, even though the unit cost of the latter system is lower. On the other hand, an architecture that offers $N/X$ MIPS at a lower unit cost would be better for the application at hand if it is possible to attain $N$ MIPS by using $Y$ such systems (where $Y = X$) with a lower total cost compared to the architecture delivering $N$ MIPS. Of course, if multiple units of an $N/X$ — MIPS machine cannot be configured to deliver $N$ MIPS, then it is not a candidate for comparison. This is obviously an over simplification, since configuring multiple machines to form a system typically requires other considerations such as partitioning of application into subtasks, reprogramming the sequential application into parallel form, overhead introduced by the communication between multiple processors, etc. These considerations are discussed later in this book.

The cost of a computer system is a composite of its software and hardware costs. The cost of hardware has fallen rapidly as the hardware technology progressed, while the software costs are steadily rising as the software complexity grew, despite the availability of sophisticated software engineering tools. If this trend continues, the cost of software would dictate the cost of the system while the hardware would come free once the software is purchased.

The cost of either hardware or software is dependent on two factors: an upfront development cost and per unit manufacturing cost. The development cost is amortized over the life of the system and distributed to each unit produced. Thus, as the number of systems produced increases the development component of the cost decreases.

The production cost characteristics of the hardware and software differ. Production of each unit of hardware requires assembly and testing and hence the cost of these operations will never be zero even if the cost of hardware components tends to be negligible. In the case of software, if we assume that there are no changes to the software once it is developed, resulting in zero maintenance costs, the production cost becomes almost zero as the number of units produced is large. This is because, producing a copy of the software system and testing it to make sure it is an accurate copy of the original (bit-by-bit comparison) is not an expensive operation. But, the assumption of zero maintenance costs is not realistic, since the software system always undergoes changes and enhancements are requested by the users on a continual basis.

There are other effects of progress in hardware and software technologies on the cost of the system. Each technology provides a certain level of performance and as the performance requirements increase, we exhaust the capability of a technology and hence will have to move to a new technology. Here, we are assuming that the progress in technology is user driven. In practice, the technology is also driving the users requirements in the sense that the progress in technology provides systems with higher performance at lower cost levels thereby making older systems obsolete faster than before. That means that the life spans of systems are getting shorter bringing an additional burden of recuperating development costs over a shorter period of time.

The cost considerations thus lead to the following guideline for a system architect: make the architecture as general-purpose as possible in order to make it suitable for a large number of applications, thus increasing the number of units sold and reducing the cost per unit.

## 2.8 Summary

The terminology, basic models, cost factors, and performance parameters of advanced architectures were introduced in this chapter. The most popular architecture taxonomy due to Flynn was used as the framework. A multitude of architecture taxonomies have been proposed in the literature over the years. Refer to Feng (1972), Handler (1977), Schwartz (1983), and Skillicorn (1988) for further details. Computer network model was introduced to complete the terminology. Networks are the most popular architectures today, although they are not strictly considered "high performance" structures.

## Problems

2.1. It is said that CRAY series of supercomputers can be classified under all the classes of Flynn's taxonomy. Verify this statement with respect to newer architectures in this series.

2.2. Why is the MISD classification of Flynn considered an aberration? Does a pipeline structure fit this classification?

2.3. List the characteristics of applications suitable for the following architectures:

   1. SIMD
   2. Shared memory MIMD
   3. Message passing MIMD
   4. Data-flow

2.4. How are MIMD systems different from computer networks?

2.5. MIMD systems are generally known as "multiprocessor" systems. The literature also uses the terms "multiple processor systems" and "multicomputer systems" to describe certain architectures. What is the difference between the two?

2.6. Select a parallel-processing application you are familiar with. Investigate into the algorithms available, special data structures utilized, architectural concepts used to make those algorithms efficient, and the commercial computer systems used for that application.

2.7. How does an SIMD system handle a conditional branch instruction such as IF X = 0 THEN P ELSE Q, where P and Q are labels of statements in the program? Note that in SISD, either P or Q is executed based on the value of X.

2.8. It is required to compute the sum of 32K 32-bit numbers. Estimate the time required for this computation on:

1. A 32-bit SISD
2. A 16-bit SISD
3. An SIMD with 32 32-bit processors
4. An SIMD with 16 32-bit processors

2.9. What is the best way to implement the computation of Problem (2.8) on an MIMD? That is, how do you minimize the overhead involved?

2.10. Select a processor family you have access to and estimate the speedup obtained by using a floating-point coprocessor in the system, using an appropriate benchmark program.

# References

Culler, D. E., Singh, J. P., and Gupta, A. *Parallel Computer Architecture*. San Francisco, CA: Morgan Kaufmann, 1998.

Duncan, R. A survey of parallel computer architectures. *IEEE Computer*, 23, 1990, 5–16.

Feng, T. Y. Some characteristics of associative/parallel processing. In *Proceedings of the 1972 Sagamore Computing Conference*, 5–16 August 1972.

Flynn, M. J. Very high speed computing systems. *Proceedings of the IEEE*, 54, 1966, 1901–1909.

Flynn, M. J. Some computer organizations and their effectiveness. *IEEE Transactions on Computers*, C-21, 1972, 948–960.

Foster, I. and Kesselman, C. *The GRID 2: Blueprint for a New Computing Infrastructure*. New York: Morgan Kaufmann, 2004.

Handler, W. The impact of classification schemes on computer architecture. In *Proceedings of the International Conference on Parallel Processing*, August 1977, PP. 7–15.

Kurose, J. F. and Ross, K. W. *Computer Networking: A Top-Down Approach Featuring the Internet*. Boston, MA: Addison Wesley, 2001.

Milutinovic, V. M. *High Level Language Computer Architectures*. New York: Computer Science Press, 1989.

Myers, G. J. *Advances in Computer Architecture*. New York: John Wiley & Sons, 1982.

Schwartz, J. *A Taxonomic Table of Parallel Computers Based on 55 Designs*. New York: Courant Institute, New York University, November 1983.

Skillicorn, D. B. A taxonomy for computer architectures. *IEEE Computer*, 21, 1988, 46–57.

*SPEC Benchmark Suite* Release 1.0, SPEC, Santa Clara: CA, 2000.

Treleaven, P. C., Brownbridge, D. R., and Hopkins, R. P. Data-driven and demand-driven computer architecture. *ACM Computing Surveys*, 14, 1982, 93–143.

# 3
## Pipelining

As mentioned earlier, pipelining offers an economical way of realizing parallelism in computer systems. The concept of pipelining is similar to that of an assembly line in an industrial plant wherein the task at hand is subdivided into several subtasks and each subtask is performed by a *stage (segment)* in the pipeline. In this context, *the task* is the processing performed by the conglomeration of all the stages in the pipeline, and the *subtask* is the processing done by a stage. For example, in the car assembly line described earlier, "building a car" is the task and, it was partitioned into four subtasks. The tasks are streamed into the pipeline and all the stages operate concurrently. At any given time, each stage will be performing a subtask belonging to different task. That is, if there are $N$ stages in the pipeline, $N$ different tasks will be processed simultaneously and each task will be at a different stage of processing.

The processing time required to complete a task is not reduced by the pipeline. In fact, it is increased, due to the buffering needed between the stages in the pipeline. But, since several tasks are processed simultaneously by the pipeline (in an overlapped manner), the task completion rate is higher compared to sequential processing of tasks. That is, the total processing time of a program consisting of several tasks is shorter, compared to sequential execution of tasks. As shown earlier, the throughput of an $N$-stage pipelined processor is nearly $N$ times that of the nonpipelined processor.

The next section provides a model for the pipeline and describes the types of pipelines commonly used. Section 3.2 describes pipeline control strategies. Section 3.3 deals with data interlock and other problems in pipeline design, and Section 3.4 describes dynamic pipelines. Almost all computer systems today employ pipelining techniques to one degree or another. Section 3.5 provides a selected set of examples. Chapter 4 describes the architecture of vector processors, which utilize pipeline structures extensively.

### 3.1 Pipeline Model

Figure 3.1(a) shows a pipeline with $k$ stages ($S_1, S_2, \ldots, S_k$) along with a *staging register* at the input of each stage. The input data to any stage is held in the

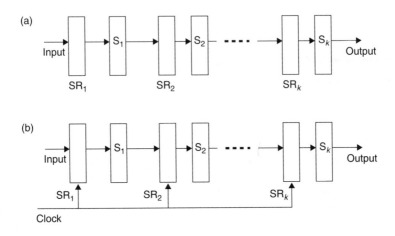

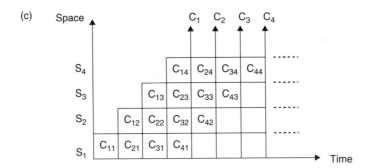

**FIGURE 3.1**
A $k$-stage pipeline: (a) asynchronous, (b) synchronous, and (c) space–time diagram

staging register until it is ready to be operated on by that stage. Let $t_i$ denote the processing time and $d_i$ denote the delay introduced by the staging register in stage $i$. Then, the total processing time $T_{pl}$ for each task through the pipeline is:

$$T_{pl} = \sum_{i=1}^{k}(t_i + d_i) \qquad (3.1)$$

The second term on the right-hand side of the above equation is the overhead introduced by the pipeline for each task because the time for sequential (i.e., nonpipelined) execution of the task is:

$$T_{seq} = \sum_{i=1}^{k}(t_i) \qquad (3.2)$$

For this pipeline to operate properly, the data produced by each stage must be input to the staging register of the following stage only after the data in that

# Pipelining

staging register has been accepted by the stage connected to it. This mode of operation results in an *asynchronous* pipeline. In general, design of asynchronous hardware is tedious, although if designed properly it provides a higher speed compared to synchronous designs. Almost all practical pipelines are designed to operate as synchronous pipelines with a clock controlling the data transfer from stage to stage as shown in Figure 3.1(b). Obviously, the frequency of the clock is $1/t_{\max}$, where $t_{\max} = \text{Max } (t_i + d_i)$, $1 \le i \le k$. The clock period is usually called the *pipeline cycle time*. The pipeline cycle time is thus controlled by the slowest stage in the pipeline.

The *space–time diagram* of Figure 3.1(c) shows the task-flow through a four-stage pipeline. Here $C_{ij}$ denotes the $j$th subtask of the task $C_i$. Note that $C_1$ is completed at the end of the fourth cycle by which the pipeline is full. From then on, one task is completed every cycle.

To attain the maximum throughput, the pipeline cycle must be as small as possible. That is, $t_{\max}$ must be minimized. This can be achieved by dividing the computational task into a large number ($k$) of subtasks, each taking approximately the same computation time (i.e., $T_{\text{seq}}/k$), resulting in a pipeline with $k$ stages and a cycle time:

$$t_{\text{cyc}} = T_{\text{seq}}/k + d \qquad (3.3)$$

where for simplicity each staging register is assumed to contribute an equal amount of delay $d$.

The smaller the value of $t_{\text{cyc}}$, the larger is the processing speed of the pipeline. But, small values for $t_{\text{cyc}}$ result in a large value for $k$. As $k$ gets large (i.e., the pipeline gets deeper), the first term on the right-hand side of (3.3) tends to zero and $t_{\text{cyc}}$ tends to $d$. Thus, the pipeline computation rate is bounded by $1/d$. As $k$ increases, the overhead in terms of filling the pipeline at the beginning of the task stream and coordination of dataflow between the stages also increases. Also, each stage in the pipeline potentially accesses the system memory for data and instructions. Since all the $k$ stages are active simultaneously, as $k$ becomes large, the memory traffic increases $k$-fold, thus making the memory system a bottleneck.

Suppose the clock-rate of the pipeline is fixed and suppose all stages of a linear pipeline are able to complete their operations in one-clock cycle, except stage X, which requires two-clock cycles. Because of stage X, the pipeline can produce at most one result every two-clock cycles. It is possible to convert the pipeline to produce one result every cycle by breaking the stage X into two units. But, this requires addition of another latch, thus increasing the total delay. As such, it may be necessary to break the stage X into more than two parts to achieve the desired goal. For pipelines where the maximum clock rate is determined by the slowest stage, it may appear to be possible to speedup the pipeline arbitrarily by dividing each stage into smaller and smaller parts. However, because the delay of the staging registers is fixed, the pipeline will eventually reach a point where the amount of improvement in speed is negligible, and the cost of making such improvements becomes prohibitive.

If $N$ tasks are streamed into the pipeline of Figure 3.1(b), the first $k$ cycles are used to fill the pipeline, at the end of which the first result appears at the output and the remaining $(N-1)$ tasks are completed in subsequent $(N-1)$ cycles, thus taking a total time of $(k+N-1)t_{cyc}$. Substituting from equation (3.3), the pipeline processing time becomes $(k+N-1)(T_{seq}/k+d)$ Thus, the staging register delay $d$ contributes an overhead of the order of $(k+N-1)$. This overhead can only be reduced either by using registers with the smallest possible delay or by limiting the number of stages $k$ in the pipeline to a small number. An ideal design thus strikes a compromise between the number of stages and the pipeline cycle time.

The execution of $N$ tasks in a nonpipelined processor would require $(N \cdot T_{seq})$ time units. The *speedup* $S$ obtained by the pipeline is defined as the ratio of the total processing time of the nonpipelined implementation to that of the pipelined implementation and is given by:

$$S = \frac{N \cdot T_{seq}}{(k+N-1)t_{cyc}} \tag{3.4}$$

If staging register delay $d$ is ignored, equation (3.4) reduces to the speedup of an ideal pipeline:

$$S_{ideal} = (N \cdot k)/(k+N-1) \tag{3.5}$$

We can include a cost factor into this model since the parameter of general interest while evaluating different pipeline architectures is the cost per million instructions per second (MIPS) or cost per million floating-point operations per second (MFLOPS). If the hardware cost of the $i$th stage is $c_i$ and the cost of each staging register is $L$, the total cost $C$ of the pipeline is given by:

$$C = L \cdot k + C_p \tag{3.6}$$

where

$$C_p = \sum_{i=1}^{k} c_i \tag{3.7}$$

The cost function $C$ increases linearly with the number of stages $k$.

The composite cost per computation rate given by $(R \cdot C_p)$. That is,

$$\begin{aligned} R \cdot C_p &= (L \cdot k + C_p)(T_{seq}/k + d) \\ &= L \cdot T_{seq} + L \cdot d \cdot k + C_p \cdot T_{seq}/k + C_p \cdot d \end{aligned} \tag{3.8}$$

# Pipelining

To minimize this,

$$\frac{d}{dk}(R \cdot C_p) = 0 \quad (3.9)$$

That is,

$$L \cdot d - C_p \cdot T_{seq}/k^2 = 0$$

or

$$k = \sqrt{(C_p \cdot T_{seq})/(L \cdot d)} \quad (3.10)$$

This provides the condition for the lowest cost per computation.

In practice, making the delays of pipeline stages equal (as assumed above), is a complicated and time consuming process. It is essential to maximum performance that the stages are close to balanced, but designing circuitry to achieve that requires analyzing a huge number of paths through the circuits and iterating a large number of times over complicated steps. It is done for commercial processors, although it is not easy or cheap to do.

Another problem with pipelines is the overhead in terms of handling exceptions or interrupts. Interrupts can be handled in two ways:

1. The contents of all the stages are saved for an orderly return to that state after the interrupt is serviced.
2. The contents of the pipeline are discarded (except may be the instruction in the last stage, which is completed before starting the interrupt service), interrupt serviced and the pipeline is restarted. Note also that a deep pipeline increases the interrupt handling overhead.

### 3.1.1 Pipeline Types

It is important to note that pipeline concept can be applied at various levels. The complete application program can be developed as several modules with each module executing on a pipeline stage; each module could be implemented as a pipeline in which each stage executes a statement or a set of statements within the module; each statement may be executed by a pipeline in which a stage performs one operation in the statement; each operation in a statement can subsequently be executed by a pipelined arithmetic/logic unit.

There are two broad types of pipelines: Instruction pipelines and Arithmetic pipelines. *Instruction pipelines* are used in almost all modern day processors to a varying degree, to enhance the speed of the control unit. That is, the control unit is implemented as a pipeline consisting of several stages, where each stage is dedicated to perform one phase of the instruction cycle. *Arithmetic pipelines* enhance the throughput of arithmetic/logic units. Further details on these two types of pipelines follow.

### 3.1.1.1   Instruction pipeline

Figure 3.2 shows an instruction processing pipeline consisting of six stages. The first stage fetches instructions from the memory, one instruction at a time. At the end of each fetch, this stage also updates the program counter to point to the next instruction in sequence. The decode stage decodes the instruction, the next stage computes the effective address of the operand, followed by a stage that fetches the operand from the memory. The operation called for by the instruction is then performed by the execute stage and the results are stored in the memory by the next stage.

The operation of this pipeline is depicted by the modified time–space diagram called a *reservation table* (RT) shown in Figure 3.3. In an RT, each row corresponds to a stage in the pipeline and each column corresponds to a pipeline cycle. An "X" at the intersection of the $i$th row and the $j$th column indicates that stage $i$ would be busy performing a subtask at cycle $j$, where

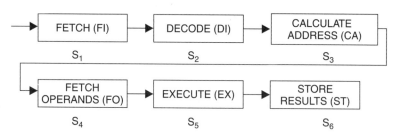

**FIGURE 3.2**
An instruction pipeline

**FIGURE 3.3**
Reservation table

# Pipelining

cycle 1 corresponds to the initiation of the task in the pipeline. That is, stage $i$ is "reserved" (and hence not available for any other task) at cycle $j$. The number of columns in the RT for a given task is determined by the sequence in which the subtasks corresponding to that task, flow in the pipeline. The RT in Figure 3.3 shows that each stage completes its task in one cycle time and hence an instruction cycle requires six cycles to be completed, although one instruction is completed every cycle (once the pipeline is full).

In the above pipeline, the cycle time is determined by the stages requiring memory access because they tend to be slower than the other stages. Assume that a memory access takes $3T$, where $T$ is a time unit and a stage requiring no memory access executes in $T$. Then, the above pipeline produces one result every $3T$ (after the first $18T$, during which the pipeline is "filled"). Compared to this, the sequential execution of each instruction requires $14T$ on asynchronous and $18T$ on synchronous control units.

The above pipeline can be rearranged into a 14-stage pipeline shown in Figure 3.4 (Stone, 1987). Here, two delay stages are inserted at each stage requiring a memory access in the original pipeline. This implies that the memory system is reorganized into a three-stage pipeline. The delay stages provide for the buffering of data to and from the memory. Now the pipeline

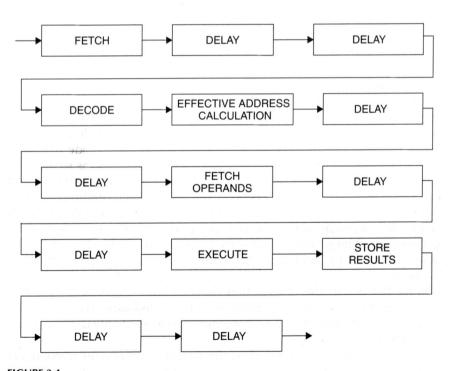

**FIGURE 3.4**
Instruction pipeline with delays

produces one result every $T$ (ignoring the additional overhead due to staging registers needed between the stages).

The assumption so far has been that the instruction execution is completely sequential in the operation of the above pipeline. As long as that is true the pipeline completes one instruction per cycle. In practice, the program execution is not completely sequential due to branch instructions. Consider an unconditional branch instruction entering the pipeline of Figure 3.2. The target of the branch is not known until the instruction reaches the address calculate stage $S_3$. By then, if the pipeline is allowed to function normally, it would have fetched two more instructions following the branch instruction. When the target address is known, the instructions that entered the pipeline after the branch instruction must be discarded and new instructions fetched from the target address of the branch. This pipeline draining operation results in a degradation of pipeline throughput. A solution would be to freeze the pipeline from fetching further instructions as soon as the branch opcode is decoded (in stage 2) until the target address is known. This mode of operation prevents some traffic on the memory system, but does not increase the pipeline efficiency any, compared to the first method.

When the instruction entering the pipeline is a conditional branch, the target address of the branch will be known only after the evaluation of the condition (in stage 5). Three modes of pipeline handling are possible for this case. In the first mode the pipeline is frozen from fetching subsequent instructions until the branch target is known, as in the case of unconditional branch above. In the second mode, the pipeline fetches subsequent instructions normally, ignoring the conditional branch. That is, pipeline predicts that the branch will not be taken. If indeed the branch is not taken, the pipeline flows normally and hence there is no degradation of performance. If the branch is taken, the pipeline must be drained and restarted at the target address. The second mode is preferred, since the pipeline functions normally about 50% of the time on an average. The third mode would be to start fetching the target instruction sequence into a buffer (as soon as the target address is computed in $S_3$), while the nonbranch sequence is being fed into the pipeline. If the branch is not taken, the pipeline continues with its normal operation and the contents of the buffer are ignored. If the branch is taken, the instructions already in the pipeline are discarded (i.e., the pipeline is flushed) and target instruction is fetched from the buffer. The advantage here is that fetching instructions from the buffer is faster than that from the memory.

In the last two modes of operation above, all the activity of the pipeline with respect to instructions entering the pipeline following the conditional branch, must be marked *temporary* and made permanent only if the branch is not taken.

The above problems introduced into the pipeline by branch instructions are called control hazards. Section 3.3 describes other mechanisms that reduce the effect of *control hazards* on pipeline performance.

# Pipelining

### 3.1.1.2 Arithmetic pipelines

The most popular arithmetic operations utilized to illustrate the operation of arithmetic pipelines in the literature are: floating-point addition and multiplication. This section follows that tradition.

*Floating-point addition*

Consider the addition of two normalized floating-point numbers:

$$A = (E_a, M_a) \quad \text{and} \quad B = (E_b, M_b)$$

to obtain the sum

$$S = (E_s, M_s)$$

where $E$ and $M$ represent the exponent and mantissa, respectively. The addition follows the steps shown below:

1. Equalize the exponents:

   If $E_a < E_b$, swap $A$ and $B$; $E_{\text{diff}} = E_a - E_b$

   Shift $M_b$ right $E_{\text{diff}}$ bits

2. Add Mantissae:

$$M_s = M_a + M_b$$
$$E_s = E_a$$

3. Normalize $M_s$ and adjust $E_s$ to reflect the number of shifts required to normalize.
4. Normalized $M_s$ might have larger number of bits than can be accommodated by the mantissa field in the representation. If so, round $M_s$.
5. If rounding causes a mantissa overflow, renormalize $M_s$ and adjust $E_s$ accordingly.

Figure 3.5 shows a five-stage pipeline configuration for the addition process given above.

The throughput of the above pipeline can be enhanced by rearranging the computations into a larger number of stages, each consuming a smaller amount of time, as shown in Figure 3.6. Here, equalizing exponents is performed using a subtract exponents stage and a shift stage that shifts mantissa appropriately. Similarly, normalizing is split into two stages. This eight-stage pipeline provides a speedup of $8/5 = 1.6$ over the pipeline of Figure 3.5.

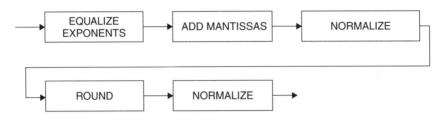

**FIGURE 3.5**
Floating-point add pipeline

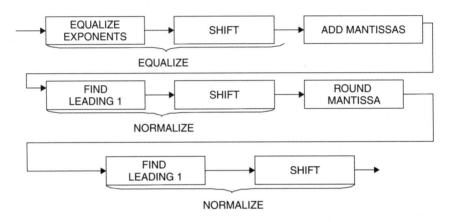

**FIGURE 3.6**
Modified floating-point add pipeline

In the pipeline of Figure 3.6 we have assumed that the shift stages can perform an arbitrary number of shifts in one cycle. If that is not the case, the shifters have to be used repeatedly. Figure 3.7 shows the rearranged pipeline where the feedback paths indicate the reuse of the corresponding stage.

Pipelines shown in Figures 3.1 through 3.6 are called *linear* pipelines, since the tasks flow from stage to stage from the input to the output. The pipeline of Figure 3.7 is *nonlinear*, because of the feedback paths it contains.

*Floating-point multiplication*

Consider the multiplication of two floating-point numbers $A = (E_a, M_a)$ and $B = (E_b, M_b)$, resulting in the product $P = (E_p, M_p)$. The multiplication follows the pipeline configuration shown in Figure 3.8 and the steps are listed below:

1. Add exponents: $E_p = E_a + E_b$.
2. Multiply mantissae: $M_p = M_a * M_b$. $M_p$ will be a double-length mantissa.
3. Normalize $M_p$ and adjust $E_p$ accordingly.

4. Convert $M_p$ into single-length mantissa by rounding.
5. If rounding causes a mantissa overflow, renormalize and adjust $E_p$ accordingly.

Stage 2 in the above pipeline would consume the largest amount of time. In Figure 3.9 stage 2 is split into two stages, one performing partial products and the other accumulating them. In fact, the operations of these two stages can be overlapped in the sense that when the accumulate stage is adding, the other stage can be producing the next partial product.

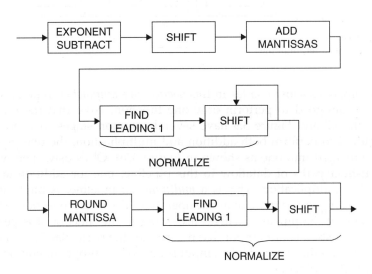

**FIGURE 3.7**
Floating-point adder with feedback

**FIGURE 3.8**
Floating-point multiplication pipeline

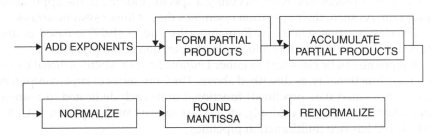

**FIGURE 3.9**
Floating-point multiplier pipeline with feedback loops

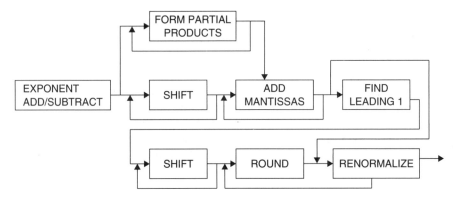

**FIGURE 3.10**
Floating-point adder/multiplier

The pipelines shown so far in this section are *unifunction* pipelines since they are designed to perform only one function. Note that the pipelines of Figure 3.7 and Figure 3.9 have several common stages. If a processor is required to perform both addition and multiplication, the two pipelines can be merged into one as shown in Figure 3.10. Obviously, there will be two distinct paths of dataflow in this pipeline, one for addition and the other for multiplication. This is a *multifunction* pipeline. A multifunction pipeline can perform more than one operation. The interconnection between the stages of the pipeline changes according to the function it is performing. Obviously, a control input that determines the particular function to be performed on the operand being input is needed for proper operation of the multifunction pipeline.

*Other pipeline types*

A brief description of other popular pipeline classification schemes is provided below.

Handler (1977) classifies pipelines into three classes. The first two classes (Instruction pipelines and Arithmetic pipelines) are similar to those described above. The third class is that of *Processor pipelining*. This class corresponds to a cascade of processors, each executing a specific module in the application program. As such, their operation resembles that of linear systolic arrays.

The classification scheme by Ramamoorthy and Li (1977) is based on the functionality, configuration, and mode of operation of the pipeline.

A pipeline can be classified as either *Unifunctional* or *Multifunctional* based on their functionality, as described above. The Cray series of supercomputers have segmented (i.e., pipelined) functional units each dedicated to a single operation (add, divide, multiply, etc.). The Control Data Corporation (CDC) STAR-100 has two multifunction pipelines.

Based on the configuration (i.e., the interconnection pattern between its stages) a pipeline is either *static* or *dynamic*. A static pipeline assumes only one functional configuration at a time and hence it is unifunctional until that configuration changes. This type of pipeline is useful when instructions of the same type can be streamed for execution. A static pipeline can be multifunctional, except that its configuration has to change for each function. Such configuration changes are very infrequent in static pipelines. That is, typically a long stream of data is processed before the configuration is changed. A dynamic pipeline on the other hand, allows more frequent changes in its configuration. The configuration might change for each input. Obviously, dynamic pipelines require more elaborate sequencing and control mechanisms compared with static pipelines.

Based on the mode of operation, a pipeline is either a *scalar* or a *vector* pipeline. A scalar pipeline is designed to process a stream of scalar operands. A loop construct normally controls the operation of this type of pipeline, sending the stream of operands from the memory into the pipeline. Vector pipelines on the other hand are used in building arithmetic logic units (ALUs) for vector processors, machines designed to execute vector instructions efficiently (see Chapter 4). Unlike the scalar data stream, the operand stream for a vector pipeline consists of elements of the same vector (array).

The following section details the pipeline control mechanisms that reduce control hazards and enhance pipeline performance.

## 3.2 Pipeline Control and Performance

For a pipeline to provide the maximum possible throughput, it must be kept full and flowing smoothly. There are two conditions that can affect the smooth flow of a pipeline: (1) the rate of input of data into the pipeline and (2) the data interlocks between the stages of the pipeline when one stage either requires the data produced by the other or has to wait until another stage needing the shared data item completes its operation on that data item. The pipeline control strategy should provide solutions for smooth flow of the pipeline under both of these conditions. The description of data interlock problems is deferred to Section 3.3. This section provides the details of a control strategy that regulates the data input rate.

**Example 3.1**
Consider the RT for the six-stage pipeline of Figure 3.3. This RT implies a strict linear operation of the pipeline. That is, each stage completes its work in one cycle and there are no feed back paths in the pipeline. As such, a new set of operands can be input to the pipeline every cycle and correspondingly the pipelines completes one operation per cycle (once it is full).

**Example 3.2**
Consider the nonlinear pipeline shown in Figure 3.11(a). The last two stages in the pipeline are used twice by each operand. That is, stage 3 produces a partial result and passes it on to stage 4. While stage 4 is processing the partial result, stage 3 produces the remaining part of the result and passes it to stage 4. Stage 4 passes the complete result to stage 2. The operation of this pipeline is represented by the RT shown in Figure 3.11(b). Since stages 3 and 4 are used in subsequent cycles by the same set of operands, the operand input rate cannot be as fast as one per cycle. The RT can be used to determine the maximum input rate of such pipelines as described next.

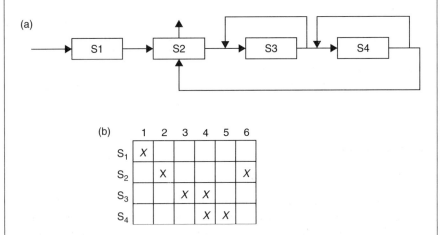

**FIGURE 3.11**
A pipeline with feedback: (a) pipeline and (b) reservation table

The overlapped operation of stages (as in the pipeline of the above example) is primarily due to the nonavailability of appropriate hardware and hence is termed *structural hazard*. For instance, only a single-bit shifter may be available and a multiple-bit shift capability is required for the task at hand; or a task requires an addition followed by a subtraction and only one ALU is available. One obvious way of avoiding structural hazards is to insert additional hardware into the pipeline. This increases the pipeline complexity and may not be justified especially if the operation requiring additional hardware is infrequent. The following example illustrates the concept of structural hazards with respect to memory requirements.

So far we have concentrated on the operation of pipelines for a single instruction cycle or a single arithmetic operation. The following example illustrates the operation of a pipeline for an instruction sequence (i.e., program).

# Pipelining

> **Example 3.3**
> Consider now, the instruction sequence given below, to be executed on a processor that utilizes the pipeline of Figure 3.2:
>
> LOAD R1, MEM1      R1 ← (MEM1)
> LOAD R2, MEM2      R2 ← (MEM2)
> MPY R3, R1         R3 ← (R3) ∗ (R1)
> ADD R4, R2         R4 ← (R4) + (R2)
>
> where R1, R2, R3, and R4 are registers, MEM1 and MEM2 are memory addresses, "( )" denote "contents of" and "←" indicates a data transfer.

Figure 3.12 depicts the operation of the pipeline. During cycles 1 and 2 only one memory access is needed. In cycle 3, two simultaneous read accesses to memory are needed, one due to CA and the other due to FI. In cycle 4, three memory read accesses (FO, CA, and FI) are needed. In cycles 5, two memory read (FO, CA) and one register write (EX) are required. Cycle 6 requires one memory read access and register read and write. Cycles 7, 8, and 9 do not require memory access. Cycle 7 requires one register read and one register write and cycle 8 needs one register write. The memory and register accesses have been summarized in Table 3.1.

As seen from Table 3.1, the memory system must accommodate three accesses per cycle for this pipeline to operate properly. If we assume that the machine has separate data and instruction caches, then two simultaneous accesses can be handled. This solves the problem in cycles 3 and 5 (assuming that the machine accesses instruction cache during CA). But, during cycle 4 two accesses (FI, CA) to data cache would be needed. One way to solve this problem is to *stall* the ADD instruction (i.e., initiate ADD instruction later than

|  | \multicolumn{9}{c}{Cycles} |
|---|---|---|---|---|---|---|---|---|---|
|  | 1 | 2 | 3 | 4 | 5 | 6 | 7 | 8 | 9 |
| Load R1, Mem1 | FI | DI | CA | FO<br>Read Mem1 | EX<br>Write R1 | ST |  |  |  |
| Load R2, Mem2 |  | FI | DI | CA | FO<br>Read Mem2 | EX<br>Write R2 | ST |  |  |
| Mpy R3, R1 |  |  | FI | DI | CA | FO<br>Read R1,R3 | EX<br>Write R3 | ST |  |
| Add R4, R2 |  |  |  | FI | DI | CA | FO<br>Read R2,R4 | EX<br>Write R4 | ST |

**FIGURE 3.12**
Pipeline operation and resources requirements

**TABLE 3.1**

Memory and register access (read and write)

| Cycle | Memory read | Memory write | Register read | Register write |
|---|---|---|---|---|
| 1 | 1 | | | |
| 2 | 1 | | | |
| 3 | 2 | | | |
| 4 | 3 | | | |
| 5 | 2 | | | 1 |
| 6 | 1 | | 1 | 1 |
| 7 | | | 1 | 1 |
| 8 | | | | 1 |
| 9 | | | | |

|  | Cycles | | | | | | | | |
|---|---|---|---|---|---|---|---|---|---|
|  | 1 | 2 | 3 | 4 | 5 | 6 | 7 | 8 | 9 |
| Load R1, Mem1 | FI | DI | CA | FO<br>Read Mem1 | EX<br>Write R1 | ST | | | |
| Load R2, Mem2 | | FI | DI | CA | FO<br>Read Mem2 | EX<br>Write R2 | ST | | |
| Mpy R3, R1 | | | FI | DI | CA | FO<br>Read R1,R3 | EX<br>Write R3 | ST | |
| Add R4, R2 | | | | Stall | Stall | FI | DI | CA | FO<br>Read R2,R4 |

**FIGURE 3.13**
Pipeline operation with stalls

cycle 4) until cycle 6 as shown in Figure 3.13. Register read operations move to cycle 9 and the register write to cycle 10 (not shown in the figure) The stalling process results in a degradation of pipeline performance. The memory and register accesses for Figure 3.13 have been summarized in Table 3.2.

Note that the pipeline controller must evaluate the resource requirements of each instruction before the instruction enters the pipeline, so that structural hazards are eliminated. The following section describes one such mechanism.

### 3.2.1 Collision Vectors

In what follows, *initiation* refers to the launching of an operation into the pipeline. This corresponds to inputting a new set of data into a static or unifunction pipeline, or data and operation designator into a dynamic or multifunction pipeline. The number of cycles that elapse between two

# Pipelining

**TABLE 3.2**

Memory and register access (read and write) for pipeline operation with stalls

| Cycle | Memory read | Memory write | Register read | Register write |
|---|---|---|---|---|
| 1 | 1 | | | |
| 2 | 1 | | | |
| 3 | 2 | | | |
| 4 | 2 | | | |
| 5 | 2 | | | 1 |
| 6 | 1 | | 1 | 1 |
| 7 | | | | 1 |
| 8 | | 1 | | |
| 9 | | | 1 | |

initiations is referred to as the *latency*. A *latency sequence* denotes the latencies between successive initiations. A latency cycle is a latency sequence that repeats itself. A *collision* occurs if a stage in the pipeline is required to perform more than one task at any time.

The RT shows the busy/nonbusy status of each pipeline stage at a given time slot (cycle) subsequent to initiation of an operation at the first cycle. In order to determine if there will be a collision at any stage at any time if another operation is initiated during the second cycle, a copy of the RT should be superimposed but shifted right one cycle, over the original RT.

### Example 3.4

Figure 3.14(a) shows such a superimposed RT corresponding to the pipeline of Figure 3.11. Here 'A' designates the reservation of stages due to the first initiation and 'B' designates that due to the second initiation. Since there are two entries in at least one block of the composite RT implying a collision, we cannot initiate an operation in the second cycle.

This superimposition of the RT can be continued to test if an operand input is possible at subsequent times after the first initiation, by simply shifting the superimposing copy of the RT by appropriate time slots.

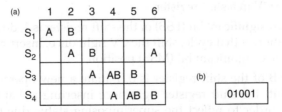

**FIGURE 3.14**
CV computation: (a) overlapped reservation table and (b) collision vector

## Example 3.5
Figure 3.14(b) shows a *collision vector* (CV) derived by repeated shift and superimposition of RT of Figure 3.11(b) for five time slots following the first input. In this vector the bits are numbered starting with 1 from right. If the $i$th bit is 1, there will be a collision in the pipeline if an operation is initiated at the $i$th cycle subsequent to an initiation. If the $i$th bit is 0, there is no collision. For the above RT, a new initiation is always possible at the sixth cycle after any initiation. Thus, the number of bits in the CV is always one less than the number of columns in the RT. This CV implies that a new operation cannot be initiated either during the first or the fourth cycle after the first initiation.

A collision occurs if two operations are initiated with latency equal to the column distances between two entries on some row of the RT. The set of all possible column distances between pairs of entries in each row of a RT is called the *forbidden set* $F$ of latencies. The CV can thus be derived from $F$ as follows:

$$CV = (v_{n-1}, v_{n-2}, \ldots, v_2, v_1)$$

where $v_i$ is 1 if $i$ is in $F$ and $v_i$ is 0 otherwise, and $n$ is the number of columns in the RT.

## Example 3.6
In the RT of Figure 3.11(b), the forbidden set $F$ is {1, 4} because third and fourth rows have pairs of entries that are 1 column apart and row 2 has entries that are 4 columns apart. Thus the CV is (01001).

### 3.2.2 Control

The CV can be utilized to control the initiation of operations in the pipeline.

1. Place the CV in a shift register.
2. If the least significant bit (LSB) of the shift register is 1, do not initiate an operation at that cycle; shift the CV right once, inserting a 0 at the vacant most significant bit (MSB) position.
3. If the LSB of the shift register is 0, initiate a new operation at that cycle; shift the shift register right once inserting a 0 at the vacant MSB. In order to reflect the superimposing status due to the new initiation over the original one, perform a bit-by-bit OR the original CV with the (shifted) content of the shift register.

# Pipelining

The insertion of 0s at the MSB insures that an operation is guaranteed an initiation within a time no longer than the length of CV, after the first initiation.

### 3.2.3 Performance

Figure 3.15(a) shows the state transitions (indicated by the contents of control shift register) of a four-stage pipeline with the CV (00111). The numbers on the arcs correspond to the cycle number at which the transition takes place, counting from the previous state. Until the pipeline reaches state 4 no operands can be input into the pipeline. After the fourth state (cycle) from the initiation of an operation, another operation can be initiated, taking the pipeline back to the original state. If the second operation is not initiated at state 4, the pipeline reaches state 5. Similarly, if the operation is initiated at state 5, the pipeline reaches state 1, otherwise it reaches state 6, after which a new operation can always be initiated.

In the determination of the performance of the pipeline, states 2, 3, and 4 are not of importance since they correspond to the conditions not allowing an initiation. The state diagram of Figure 3.15(a) can thus be simplified

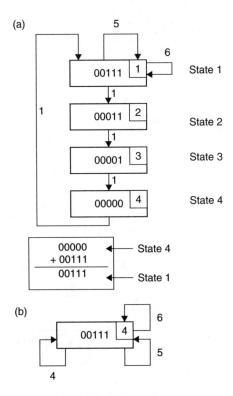

**FIGURE 3.15**
State transition of a four-stage pipeline: (a) state diagram and (b) reduced state diagram

to the reduced state diagram shown in Figure 3.15(b). Here, the transitions corresponding to only the zero bits of CV are shown. The number on the arc between two states is the latency between two initiations represented by those states and each state is represented by the content of the control shift register after the initiation of the new operation from the previous state. Thus, for this pipeline, the reduced state diagram indicates the three possible initiating conditions. As can be seen, the maximum input rate possible is 1 every four cycles. We can thus utilize the reduced state diagram to measure the performance of the pipeline. We will illustrate this with another example.

**Example 3.7**
Consider the RT of a four-stage pipeline shown in Figure 3.16(a). The CV for this RT is (101010). The reduced state diagram is shown in Figure 3.16(b). State 1 is the initial state at which an operation is initiated. As indicated by the CV, a new operation can be initiated at

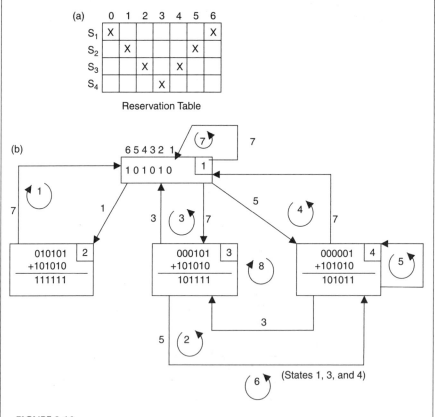

**FIGURE 3.16**
State transition example: (a) reservation table and (b) reduced state diagram

# Pipelining

latencies 1, 3, and 5 ( corresponding to the zero bits of CV) while in this state, reaching states 2, 3, and 4, respectively. From each of these new states, possible transitions are indicated in the diagram along with their latencies. Since no new states are reached, the diagram is complete.

Consider the loop (loop 1) formed by states 1 and 2 in the above reduced state diagram. If the pipeline were to follow this loop, it would generate 2 outputs (once it is full) for every $1+7 = 8$ cycles. That is, the number of states traversed by the loop indicates the number of outputs from the pipeline (also, the number of operations that can be initiated) in a time frame obtained by adding the latencies on the arcs forming the loop. That is, the *average latency* for the pipeline traversing this loop is 8/2 or 4. Similarly loop 2, provides an average latency of 4. The steady-state output rate of the pipeline is the reciprocal of the average latency. Thus, this loop provides 0.25 outputs every pipeline cycle. In order for the pipeline to follow the pattern indicated by this loop, the pipeline controller has to bring the pipeline to state 3 from the initial state (state 1) and the initiate operations at 5 and 3 cycle intervals, respectively. The loops in the reduced state diagram are usually termed "cycles." A *simple cycle* is a closed loop in which each state appears only once per traversal through the cycle. Table 3.3 lists all simple cycles and their average latencies from the above state diagram.

In order to determine the maximum performance possible, all simple cycles in the reduced state diagram need to be evaluated to select the one that provides the largest number of outputs per cycle. The *greedy cycles* in the state diagram are first candidates for such an evaluation. A greedy cycle is a simple cycle in which each latency is the minimal latency from any state in the cycle. Cycles 1 and 2 above are greedy cycles and the maximum performance of the above pipeline is two operations every eight cycles or 0.25 operations/pipeline cycle. Note in general that a greedy cycle need not be the maximum performance cycle.

**TABLE 3.3**

Average latencies

| Cycle | Avg. latency |
|---|---|
| 1[a] | 4 |
| 2[a] | 4 |
| 3 | 5 |
| 4 | 6 |
| 5 | 5 |
| 6 | 5 |
| 7 | 7 |
| 8 | 5 |

[a]Indicates "Greedy cycles."

94                                                                                              Advanced Computer Architectures

The bound on the highest possible initiation rate of a pipeline can be obtained by an examination of the RT. The row of the RT with the maximum number of entries corresponds to the bottleneck in the pipeline and limits the initiation rate. If there are Y marks in this row, the average delay between initiations (i.e., Minimum Average Latency or MAL) must be Y or greater and hence the maximum attainable rate in the pipeline is $N/Y$, where $N$ is the number of stages in the pipeline. Note that the MAL is the smallest average latency that can be achieved by any permissible latency sequence.

In the RT of Figure 3.13, the maximum number of entries in any row is 2. Thus, the MAL is 2/6 or 0.33. The above analysis predicts the maximum performance of 0.25, which is lower than the theoretical upper bound of 0.33.

The above observation on MAL is due to the following Lemma due to Shar (1972): For any statically configured pipeline executing some RT, the MAL is always greater than or equal to the maximum number of marks in any single row of the RT.

Assume that the initiation rate over a long period of time to be $r$ initiations per time unit and the number of marks in the $i$th row of the RT is $N_i$. Then the utilization of the $i$th stage is $rN_i$. Since the utilization cannot be more than 100%, $rN_i$ must be less than 1 for all $i$, thus, $r$ cannot be greater than $1/\text{MAX}(Ni)$. The average latency being the reciprocal of $r$, thus cannot be greater than MAX $(N_i)$. This lower bound provides a quick estimate of the maximum performance possible for a given pipeline and the RT.

It is possible to modify an RT to achieve maximum performance by inserting delays. Refer to the book by Kogge (1981) for details.

### 3.2.4 Multifunction Pipelines

The above control strategy and performance measurement scheme can be extended to multifunction pipelines. In general, a $k$-function pipeline requires $k$ shift registers in its controller. There will be $k$ RTs each describing the stage allocation pattern for one of the $k$ operations possible. If X and Y designate two operations possible in a multifunction pipeline, $V_{xy}$ denotes a cross-CV corresponding to allowing the operation Y after X has been initiated and $V_{xx}$ denotes the CV corresponding to allowing X after X has been initiated. Thus, there will be a total of $k^2$ CVs, $k$ of which are of the type $V_{xx}$ and the remaining are of the type $V_{xy}$.

---

**Example 3.8**
Consider the RT of Figure 3.17 for a pipeline designed to perform two operations X and Y. The four CVs are shown in Figure 3.17(b). The value of each bit in $V_{xx}$ (1010) and $V_{yy}$ (0110) are determined as before for unifunction pipelines. To determine the cross-CV $V_{xy}$, the RT for operations X and Y are overlaid. Bit $i$ of the CV is 1 if in some row of the overlaid

RT, column $t$ contains an X and column $(t + i)$ contains a Y; otherwise it is 0. Thus, $V_{xy} = (1001)$ and $V_{yx} = (0101)$. In Figure 3.17(c), the CVs are grouped into two collision matrices where the collision matrix **MX** is the collection of CVs corresponding to initiating X after some other operation has been initiated (i.e., X is the second subscript of the CV).

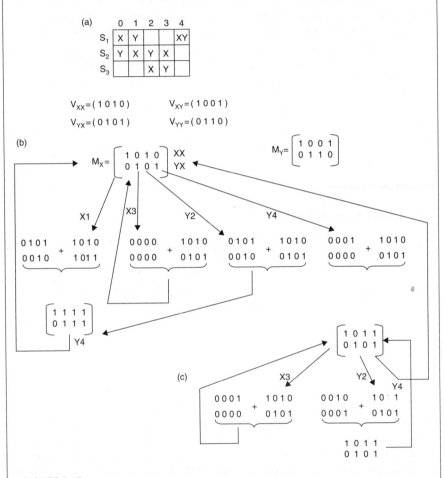

**FIGURE 3.17**
Multifunction pipeline: (a) reservation table, (b) collision vectors, and (c) reduced state diagram (partial)

The control strategy for the multifunction pipeline is similar to that for the unifunction pipeline. The collision matrix $\mathbf{M_X}$ controls the insertion of the operation X. The controller requires $k$ shift registers. To determine the insertion of operation X at any cycle, $\mathbf{M_X}$ is loaded into $k$ control shift registers. Insertion of X after the initiation of operation $i$ is determined by the shift register $i$ (i.e., $Vi_X$). If the rightmost bit of shift register $i$ is 0 operations X is allowed, otherwise it is not allowed to enter the pipeline. All the shift registers are shifted right at each cycle with a 0 entering each shift register at the left. To determine the new state, the shifted matrix is bitwise ORed with the original collision matrix. The reduced state diagram is drawn starting with k initial states, each corresponding to a collision matrix.

Figure 3.17(c) shows the reduced state diagram. Initial state $S_X$ corresponds to $\mathbf{M}_X$ and $S_Y$ corresponds to $\mathbf{M}_Y$. There are two components in the marks on each arc leaving a state $S_P$: the first component designates a row of $\mathbf{M}_P$ and the second component designates a zero-bit position of that row. For example, X1 on the arc leaving $\mathbf{M}_X$ corresponds to the insertion of X, in the first cycle after X has been initiated; Y2 corresponds to the insertion of X two cycles after the initiation of Y, and so forth.

The method for finding the greedy cycles and the maximum performance of multifunction pipelines is complex although it is similar to that for unifunction pipelines (Thomas, 1978).

## 3.3 Other Pipeline Problems

In addition to the collision problems in pipelines due to improper initiation rate, data interlocks occur due to shared data between the stages of the pipeline and conditional branch instructions degrade the performance of an instruction pipeline, as mentioned earlier. These problems and some common solutions are described in this section.

### 3.3.1 Data Interlocks

An instruction processing pipeline is most efficient when instructions flow through its stages in a smooth manner. In practice, this is not always possible because of the inter-instruction dependencies. These inter-instruction dependencies are due to the sharing of resources such as a memory location or a register by the instructions in the pipeline. In such sharing environments, the computation cannot proceed if one of the stages is operating on the resource while the other has to wait for the completion of that operation.

---

**Example 3.9**
Consider the following instruction sequence:

| | |
|---|---|
| LOAD R1, MEM1 | R1 ← (MEM1) |
| LOAD R2, MEM2 | R2 ← (MEM2) |
| MPY R1, R2 | R1 ← (R2) * (R1) |
| ADD R1, R2 | R1 ← (R1) + (R2) |

# Pipelining

Figure 3.18 shows the operation of the pipeline of Figure 3.2 for this sequence. Note that as a result of the second LOAD instruction, R2 is loaded with the data from MEM2 during cycle 6. But, the MPY instruction reads R2 during cycle 6 also. In general R2 cannot be guaranteed to contain the proper data until the end of cycle 6 and hence MPY instruction would operate with erroneous data. Similar *data hazard* occurs in cycle 7. For the results to be correct, we must ensure that R2 and R1 are read after they have been written into by the previous instruction, in each of these cycles. One possible solution is to *forward* the data to where it is needed in the pipeline as early as possible. For instance, since the ALU requires contents of R2 in cycle 6, the memory read mechanism can simply forward the data to ALU while it is being written into R2, thus accomplishing the write and read simultaneously. The concept of internal forwarding is described further, later in this section.

|  | 1 | 2 | 3 | 4 | 5 | 6 | 7 | 8 | 9 |
|---|---|---|---|---|---|---|---|---|---|
| Load R1, Mem1 | FI | DI | CA<br>Mem 1 | FO<br>Read Mem1 | EX<br>Write R1 | ST |  |  |  |
| Load R2, Mem2 |  | FI | DI | CA<br>Mem 2 | FO<br>Read Mem2 | EX<br>Write R2 | ST |  |  |
| Mpy R2, R1 |  |  | FI | DI | CA | FO<br>Read R1,R2 | EX<br>Write R1 | ST |  |
| Add R1, R2 |  |  |  | FI | DI | CA | FO<br>Read R1,R2 | EX<br>Write R1 | ST |

**FIGURE 3.18**
Data hazards

In general, the following scenarios are possible among instructions *I* and *J* where *J* follows *I* in the program:

1. Instruction *I* produce a result, which is required by *J*. Then, *J* has to be delayed until *I* produces the result.
2. Both *I* and *J* are required to write into a common memory location or a register, but the order of writing might get reversed due to the operation of the pipeline.
3. *J* writes into a register whose previous contents must be read by *I*. Then *J* must be delayed until the register contents are read by *I*.

If the order of operations is reversed by the pipeline from what was implied by the instruction sequence in the program, then the result will be erroneous.

Since, an instruction either READs from a resource or WRITEs into it, there are four possible orders of operations by two instructions that are sharing that resource. They are:

1. READ/READ (READ after READ)
2. READ/WRITE (READ after WRITE)
3. WRITE/READ (WRITE after READ)
4. WRITE/WRITE (WRITE after WRITE)

In each case, the first operation is from the earlier instruction $I$ and the second operation is from the later instruction $J$. If the orders are reversed, a *conflict* occurs. That is, a READ/WRITE conflict occurs if the WRITE operation is performed by $J$ before the resource has been READ by $I$, and so on.

Reversing the order of READ/READ is not detrimental since data is not changed by either instructions and hence it is not considered a conflict. After a WRITE/WRITE conflict, the result in the shared resource is the wrong one for subsequent read operations. If it can be established that there are no READs in between the two WRITEs, the pipeline can allow the WRITE from $J$ and disable the WRITE from $I$ when it occurs. If the order of either READ/WRITE or WRITE/READ is reversed, the instruction reading the data gets an erroneous value. These conflicts must be detected by the pipeline mechanism to make sure that results of instruction execution remain as specified by the program.

There are in general two approaches to resolve conflicts. The first one is to compare the resources required by the instruction entering the pipeline with those of the instructions that are already in the pipeline and stall (i.e., delay the initiation) the entering instruction if a conflict is expected. That is, in the instruction sequence $[I, I+1, \ldots, J, J+1, \ldots]$, if a conflict is discovered between the instruction $J$ entering the pipeline with instruction $I$ in the pipeline, then the execution of instructions $J, J+1, \ldots$ is stopped until $I$ passes the conflict point. The second approach is to allow the instructions $J, J+1, \ldots$ to enter the pipeline and handle the conflict resolution at each potential stage where the conflict might occur. That is, suspend only instruction $J$ and allow $J+1, J+2, \ldots$ to continue. Of course, suspending $J$ and allowing subsequent instructions to continue might result in further conflicts. Thus, a multilevel conflict resolution mechanism may be needed making the pipeline control very complex. The second approach known as *instruction deferral*, may offer better performance although it requires more complex hardware and independent functional units. Section 3.4 describes instruction deferral further.

One approach to avoid WRITE/READ conflicts is *data forwarding* in which the instruction that WRITEs the data, also forwards a copy of the data to those instructions waiting for it. A generalization of this technique is the concept of *Internal forwarding* which is described next.

### 3.3.1.1 Internal forwarding

Internal forwarding is a technique to replace unnecessary memory accesses by register-to-register transfers, during a sequence of read-operate-write operations on the data in the memory. This results in a higher throughput since slow memory accesses are replaced by faster register-to-register operations. This scheme also resolves some data interlocks between the pipeline stages.

Consider the memory location $M$ with which registers $r1$ and $r2$ exchange data. There are three possibilities of interest:

*write–read forwarding*: The following sequence of two operations:

$$M \leftarrow (r1)$$
$$r2 \leftarrow (M)$$

where "$\leftarrow$" designates a data transfer and "( )" designate the "contents of," can be replaced by

$$M \leftarrow (r1)$$
$$r2 \leftarrow (r1)$$

thus saving one memory access.

*read–read forwarding*: The following sequence of two operations:

$$r1 \leftarrow (M)$$
$$r2 \leftarrow (M)$$

can be replaced by

$$r1 \leftarrow (M)$$
$$r2 \leftarrow (r1)$$

thus saving one memory access.

*write–write forwarding (overwriting)*: The following sequence of two operations:

$$M \leftarrow (r1)$$
$$M \leftarrow (r2)$$

can be replaced by

$$M \leftarrow (r2)$$

thus saving one memory access.

The internal forwarding technique can be applied to a sequence of operations as shown by the following example.

**Example 3.10**
Consider the operation $P = (A * B) + (C * D)$ where P, A, B, C, and D are memory operands. This can be performed by the following sequence:

R1 ← (A)
R2 ← (R1) * (B)
R3 ← (C)
R4 ← (R3) * (D)
P ← (R4) + (R2)

The dataflow sequence for these operations is shown in Figure 3.19(a). By internal forwarding, the dataflow sequence can be altered to that in Figure 3.19(b). Here, A and C are forwarded to the corresponding multiply units eliminating register R1 and R3, respectively. The results from these multiply units are forwarded to the adder eliminating the transfers to R2 and R4.

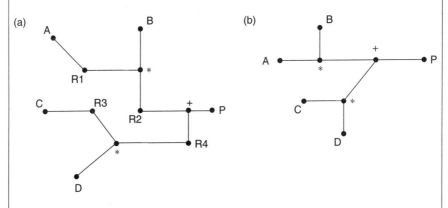

**FIGURE 3.19**
Example of internal forwarding: (a) without forwarding and (b) with forwarding

In this example, a generalized architecture that allows operations between memory and register operands (as in second and fourth instructions) and register operands (as in the last instruction) was assumed. There are two other possibilities: *load/store* and *memory/memory* architectures. In load/store architecture, the arithmetic operations are always between two register operands, and memory is used for load and store only. In memory/memory architectures, operations can be performed on two memory operands directly.

# Pipelining

Assessment of the performance of the above instruction sequence on these architectures is left as an exercise.

The internal forwarding technique as described above is not restricted to pipelines alone but is applicable to general multiple processor architectures. In particular, internal forwarding in a pipeline is a mechanism to supply data produced by one stage, to another stage that needs them, directly (i.e., without storing them in and reading them from the memory). The following example illustrates the technique for a pipeline.

---

**Example 3.11**
Consider the computation of the sum of the elements of an array $A$. That is,

$$\text{SUM} = \sum_{i=1}^{n} A_i$$

Figure 3.20(a) shows a pipeline for computing this SUM. It is assumed that the array elements and the SUM are located in memory. Assuming that each stage in the pipeline completes its task in one pipeline cycle time, computation of the first accumulation would be possible. Then onwards, the fetch sum unit has to wait for two cycles before it can obtain the proper value for the SUM. This data interlock results in the degradation of the throughput of the pipeline to one output every three cycles.

One obvious thing to do is to feed the output of the add unit back to its input as shown in Figure 3.20(b). This requires the buffering of the intermediate sum value in the add unit. Once all the elements are accumulated, the value of SUM can be stored into the memory.

If the add unit requires more than one cycle to compute the sum, the above solution again results in the degradation of throughput since the adder becomes the bottleneck.

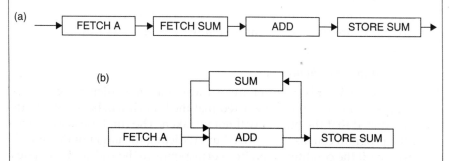

**FIGURE 3.20**
Pipeline for computing array-sum: (a) without forwarding and (b) with forwarding

Kogge (1981) provided a solution for such problems by rewriting the summing above as

$$\text{SUM}_i = \text{SUM}_{i-d} + A_i$$

where $i$ is the current iteration index and $\text{SUM}_{i-d}$ is the intermediate SUM, $d$ iterations ago and $d$ is the number of cycles required by the add unit. Because $\text{SUM}_{i-d}$ is available at the $i$th iteration, this computation can proceed every iteration. But the computation results in $d$ partial sums, each accumulating elements of $A$, $d$ apart. At the end, these $d$ partial sums should be added to obtain the final SUM. Thus, the pipeline can work efficiently at least during the computation of partial sums.

Thus in the pipeline (a) above if $d = 2$, we will require storing of two partial sums, SUM-1 and SUM-2 obtained one and two cycles ago, respectively from the current iteration. The buffer holding these partial sums should be arranged as a first-in-first-out buffer so that the fetch sum unit fetches the appropriate value for each iteration.

This type of solution is practical when changing the order of computation does not matter as in associative and commutative operations such as addition and multiplication. Even in these, changing the order might result in unexpected errors. For instance, in many numerical analysis applications the relative magnitudes of the numbers to be added are arranged to be similar. If the order of addition is changed, this structure would change resulting in a large number added to a small number, thereby altering the error characteristics of the computation.

### 3.3.2 Conditional Branches

As described earlier in this chapter, conditional branches degrade the performance of an instruction pipeline. The hardware mechanisms described earlier to minimize branch penalty were static in nature in the sense that they did not take into consideration the dynamic behavior of the branch instruction during program execution. Two compiler-based static schemes and two hardware-based dynamic schemes are described below.

#### 3.3.2.1 *Branch prediction*

The approaches described earlier are in a way branch prediction techniques, in the sense that one of them predicted that the branch will be taken and the other predicted that the branch will not be taken. The run time characteristics of the program can also be utilized in predicting the target of the branch. For instance, if the conditional branch corresponds to the end-of-do-loop test in a Fortran program, it is safe to predict that the branch is to the beginning of the loop. This prediction would be correct every time through the loop except for the last iteration. Obviously, such predictions are performed by the compiler, which generates appropriate flags to aid prediction during program execution.

In general, the target of the branch is guessed and the execution continues along that path while marking all the results as tentative until the actual target is known. Once, the target is known, the tentative results are either made permanent or discarded. Branch prediction is very effective as long as the guesses are correct. Refer to Section 1.3.1 for a description of the Branch Prediction mechanism of Intel Itanium.

### 3.3.2.2  Delayed branching

The *delayed branching* technique is widely used in the design of microprogrammed control units. Consider the two-stage pipeline for the execution of instructions shown in Figure 3.21(a). The first stage fetches the instruction and the second executes it. The effective throughput of the pipeline is one instruction per cycle, for sequential code. When a conditional branch is encountered, the pipeline operation suffers while the pipeline fetches the target instruction as shown in Figure 3.21(b). But, if the branch instruction is interpreted as "execute the next instruction and then branch conditionally," then the pipeline can be kept busy while the target instruction is being fetched, as shown in Figure 3.21(c). This mode of operation where the pipeline executes

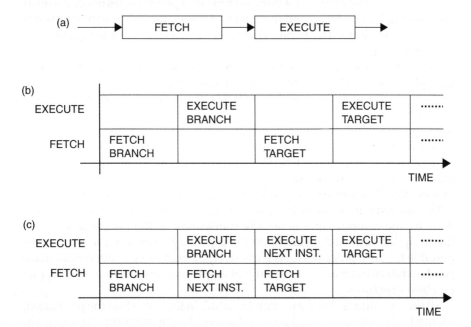

**FIGURE 3.21**
Delayed branching: (a) two-stage pipeline, (b) without delayed branching, and (c) with delayed branching

one or more instructions following a branch before executing the target of the branch is called delayed branching.

If $n$ instructions enter the pipeline after a branch instruction and before the target is known, then the branch delay slot is of length $n$ as shown below:

Branch instruction
successor instruction 1
successor instruction 2
successor instruction 3
.
.
successor instruction $n$
Branch target instruction

The compiler rearranges the program such that the branch instruction is moved $n$ instructions prior to where it normally occurs. That is, the branch slot is filled with instructions that need to be executed prior to branch and the branch executes in a delayed manner. Obviously, the instructions in the branch delay slot should not affect the condition for branch.

The length of the branch slot is the number of pipeline cycles needed to evaluate the branch condition and the target address (after the branch instruction enters the pipeline). The earlier these can be evaluated in the pipeline, the shorter is the branch slot.

As can be guessed, this technique is fairly easy to adopt for architectures that offer single-cycle execution of instructions utilizing a two-stage pipeline. In these pipelines, the branch slot can hold only one instruction. As the length of the branch slot increases, it becomes difficult to sequence instructions such that they can be properly executed while the pipeline is resolving a conditional branch.

Modern day reduced instruction set computer (RISC) architectures adopt pipelines with small number (typically 2 to 5) of stages for instruction processing. They utilize delayed branching technique extensively.

The rearrangement of instructions to accommodate delayed branching is done by the compiler and is usually transparent to the programmer. Since the compiled code is dependent on the pipeline architecture, it is not easily portable to other processors. As such, delayed branching is not considered a good architectural feature, especially in aggressive designs that use complex pipeline structures.

All the techniques above are static in nature in the sense that the predictions are made at compile time and are not changed during program execution. The following techniques utilize hardware mechanisms to dynamically predict the branch target. That is, the prediction changes if the branch changes its behavior during program execution.

### 3.3.2.3 Branch-prediction buffer

A branch-prediction buffer is a small memory buffer indexed by the branch instruction address. It contains one bit per instruction that indicates whether the branch was taken or not. The pipeline fetches subsequent instruction based on this prediction bit. If the prediction is wrong, the prediction bit is inverted. Ideally, the branch prediction buffer must be large enough to contain one bit for each branch instruction in the program, or the bit is attached to each instruction and fetched during the instruction fetch. But that increases the complexity. Typically, a small buffer indexed by several low-order bits of the branch instruction address is used to reduce the complexity. In such cases, more than one instruction maps to each bit in the buffer and hence the prediction may not be correct with respect to the branch instruction at hand, since some other instruction could have altered the prediction bit. Nevertheless, the prediction is assumed to be correct. Losq et al. (1984) named the branch-prediction buffer a *decode history table*.

The disadvantage with this technique is that by the time the instruction is decoded to detect that it is a conditional branch, other instructions would have entered the pipeline. If the branch is successful, the pipeline needs to be refilled from the target address. To minimize this effect, if the decode history table also contains the target instruction, in addition to the information as to the branch was taken or not, that instruction can enter the pipeline immediately, if the branch is a success.

### 3.3.2.4 Branch history

The *Branch history* technique (Sussenguth, 1971) uses a branch history table, which stores for each branch, the most probable target address. This target could very well be the target it reached last time during the program execution.

Figure 3.22 shows a typical branch history table. It consists of two entries for each instruction: the instruction address and the corresponding branch address. It is stored in a cache-like memory. As soon as the instruction is fetched, its address is compared with the first field of the table. If there is a match, corresponding branch address is immediately known. The execution

| Instruction address | Branch address |
|---|---|
|  |  |
|  |  |
|  |  |
|  |  |

**FIGURE 3.22**
A branch history table

continues with this assumed branch, as in branch prediction technique. The branch address field in the table is continually updated, as and when the branch targets are resolved.

As can be guessed, the above implementation of the branch history table requires an excessive number of accesses to the table, thereby creating a bottleneck at the cache containing the table.

### 3.3.3 Multiple Instruction Buffers

Section 3.1 described the use of multiple instruction buffers to reduce the effect of conditional branching on the performance of pipeline. The details of a practical architecture utilizing that feature are provided here.

Figure 3.23 shows the structure of IBM 360/91 instruction processing pipeline, where the instruction fetch stage consists of two buffers: the S-buffer is the sequential instruction prefetch buffer and the T-buffer is for the prefetch of target instruction sequence. This stage is followed by decode and other stages similar to the pipeline in Figure 3.2.

The contents of the S-buffer are invalidated when a branch is successful and the contents of T-buffer are invalidated when the branch is not successful. The decode unit fetches instructions from the appropriate buffer.

When the decoder issues a request for the next instruction the S-buffer is looked up (i.e., a single-level, non-sequential search) for sequential instructions or the T-buffer is looked up if a conditional branch has been just successful. If the instruction is available in either buffer, it is brought into the decode stage without any delay. If not, the instruction

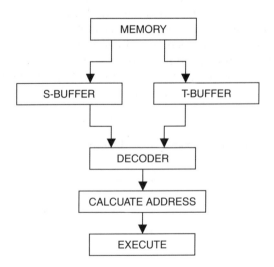

**FIGURE 3.23**
Instruction pipeline with multiple instruction buffers

needs to be fetched from the memory, which will incur a memory access delay.

All nonbranch instructions enter the remaining stages of the pipeline after the decode is complete. For unconditional branch instruction, the instruction at its target address is immediately requested by the decode unit and no further decoding is performed until that instruction arrives from the memory. For conditional branch instruction, the sequential prefetching is suspended while the instruction is traversing the remaining stages of the pipeline. Instructions are prefetched from the target address simultaneously, until the branch is resolved. If the branch is successful, further instructions are fetched from the T-buffer. If the branch is not successful, normal fetching from the S-buffer continues.

### 3.3.4 Interrupts

A complex hardware/software support is needed to handle interrupts on a pipelined processor since many instructions are executed in an overlapped manner and any of those instructions can generate an interrupt. Ideally, if instruction $I$ generates an interrupt, the execution of instructions $I+1$, $I+2, \ldots$ etc. in the pipeline should be postponed until the interrupt is serviced. However, the instructions $I-1, I-2, \ldots$ etc. that have already entered the pipeline must be completed before the interrupt service is started. Such an ideal interrupt scheme is known as *precise interrupt* scheme (Smith, 1988).

Note also that instructions are usually not processed in-order (i.e., the order in which they appear in the program). When delayed branching is used the instructions in branch-delay slot are not sequentially related. If an instruction in the branch-delay slot generates the interrupt and the branch is taken, the instructions in the branch-delay slot and the branch-target instruction must be restarted after interrupt is processed. This necessitates multiple program counters since these instructions are not sequentially related.

Several instructions can generate interrupts simultaneously and these interrupts can be out of order. That is, the interrupt due to instruction I must be handled prior to that due to $I+1$. To ensure that, a status vector is attached to each instruction as it traverses the pipeline. The machine state changes implied by the instruction are marked temporary until the last stage. At the last stage if the vector indicates that there has been no interrupt, the machine state is changed. If an interrupt is indicated, an in-order processing of interrupts is performed before state changes are committed.

## 3.4 Dynamic Pipelines

This chapter has covered static pipelines so far. In these pipelines the interconnections between the stages are fixed. A dynamic pipeline on the other hand, allows the interconnections between its stages to change (reconfigure) according to the function to be performed. It allows tasks from different RTs be initiated simultaneously. Consider the multifunction pipeline of Figure 3.10. This pipeline allows both floating-point addition and multiplication operations. But, the interconnection pattern of the stages in the pipeline is decided at the design stage for each operation and does not change as the computations proceed. When an operation is initiated in the pipeline, the controller will reserve the pipeline stages according to the pre-determined interconnection pattern for that operation.

For example, the floating-point functional unit of Texas Instruments-Advance Scientific Computer (TI-ASC) uses four ROM locations to store the interconnection patterns for each of the allowed operations. When an operation is initiated, the controller provides the address of the first location from which the stage interconnections are derived. CDC STAR-100 also allows such reconfiguration in its two static pipeline processors.

The operation of the CDC-6600 and IBM 360/91 CPUs comes close to that of a dynamic pipeline. Each of these CPUs contain multiple functional units. The functional units are not interconnected in a pipeline fashion, but communicate with each other either through registers or through a bus. The computation at hand establishes a logical interconnection pattern, which changes as the computation proceeds. The organization of these CPUs allow the implementation of instruction deferral scheme mentioned earlier.

### 3.4.1 Instruction Deferral

Instruction deferral is used to resolve the data interlock conflicts in the pipeline. The concept is to process as much of an instruction as possible at the current time and defer the completion until the data conflicts are resolved, thus obtaining a better overall performance than stalling the pipeline completely until data conflicts are resolved.

#### 3.4.1.1 CDC 6600 scoreboard (Thornton, 1970)

The processing unit of CDC 6600 shown in Figure 3.24 consists of 16 independent functional units (five for memory access, four for floating-point operations, and seven for integer operations). The input operands to each functional unit come from a pair of registers and the output is designated to one of the registers. Thus, three registers are allocated to each functional unit corresponding to each arithmetic instruction. The control unit contains a *scoreboard*, which maintains the status of all the registers, the status of functional units, and the register-functional unit associations. It contains an instruction queue called *reservation station*. Instructions first enter into the reservation station. Each arithmetic instruction corresponds to a 3-tuple consisting of two source register and one destination register designations. The load and store instructions consist of a 2-tuple corresponding to the memory address and register designation. The source operands

# Pipelining

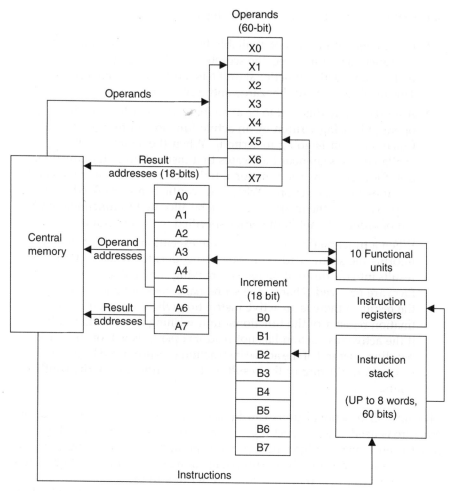

**FIGURE 3.24**
Central processor operating registers of CDC 6600 (Courtesy of Control Data Corporation.)

in each instruction have a tag associated with them. The tag provides an indication of the availability of the operand in the register. If the operand is not available, it indicates the functional unit from which the operand is expected.

For each instruction the scoreboard determines whether the instruction can be executed immediately or not based on the analysis of data dependencies. If the instruction can not execute immediately, it is placed in the reservation station and the scoreboard monitors its operand requirements, and decides when all the operands are available. The scoreboard also controls when a functional unit writes its result into the destination register. Thus, the scoreboard resolves all conflicts. The scoreboard activity with respect to each instruction can be summarized into the following steps:

1. If a functional unit is not available the instruction is stalled. When the functional unit is free, the scoreboard allocates the instruction to it, if no other active functional unit has the same destination register. This resolves structural hazards and write/write conflicts.

2. A source operand is said to be available if the register containing the operand is being written by an active functional unit or if no active functional unit is going to write it. When the source operands are available, the scoreboard allocates that instruction to the functional unit for execution. The functional unit then reads the operands and executes the instruction. Write/read conflicts are resolved by this step. Note that the instructions may be allocated to functional units out of order (i.e., not in the order specified in the program).

3. When a functional unit completes the execution of an instruction, it informs the scoreboard. The scoreboard decides when to write the results into the destination register making sure that read/write conflicts are resolved. That is, the scoreboard does not allow writing of the results if there is an active instruction whose operand is the destination register of the functional unit wishing to write its result, or if the active instruction has not read its operands yet, or the corresponding operand of the active instruction was produced by an earlier instruction. Writing of the result is stalled until read/write conflict clears.

Because all functional units can be active at any time, an elaborate bus structure is needed to connect registers and funs. The 16 functional units of the CDC 6600 were unitgrouped into four groups with a set of busses (data trunks) for each group. Only one functional unit in each group could be active at any time. Note also that all results are written to the register file and the subsequent instruction has to wait until such a write takes place. That is, there is no forwarding of the results. Thus, as long as the write/write conflicts are infrequent, the scoreboard performs well.

### 3.4.1.2  IBM 360/91 and Tomosulo's algorithm

Figure 3.25 shows the floating-point unit of IBM 360/91. It consists of four floating-point registers, a floating-point instruction queue and floating-point adder and multiplier functional units. Unlike CDC 6600, the functional units are pipelined and are capable of starting at most one operation every clock cycle. The reservation stations at each functional unit decide when an instruction can begin execution in that unit, unlike the centralized reservation station of the CDC 6600 scoreboard. The load buffers hold the data from the memory and the store buffers hold the data destined to memory. The results from functional units are broadcast to all the destinations (reservation stations, registers, and memory buffers) that require them simultaneously over the common

# Pipelining

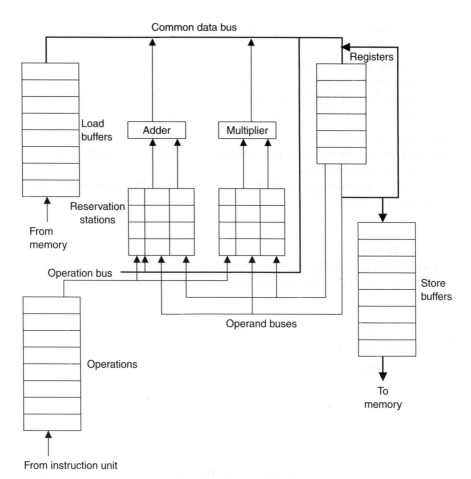

**FIGURE 3.25**
IBM 360/91 floating-point unit

data bus (CDB), rather than through registers as in CDC 6600. Instructions obtain their operands from reservation stations rather than registers, thus overcoming the limitation of small number of registers.

The following steps constitute Tomosulo's algorithm (1967) for instruction execution:

1. Get the floating-point instruction from the queue and allocate it to an appropriate reservation station if one available and also send operands from registers (if they are there) to that reservation station. The tag field at the reservation station for each operand indicates whether the operand is available or it contains an indication of which memory buffer it may be in or which functional unit is producing it. The memory buffers also have a similar tag. If the instruction is a load or store, it is allocated if an empty buffer is available. The instruction

stalls if either the memory buffer is not free or reservation station is not available.

2. Execute the instruction if both the operands are available and send the result to the CDB. If an operand is not available in the reservation station, monitor CDB until that operand is computed by another unit. Write/read conflicts are resolved in this step.

These steps are similar to that in the scoreboard except that the write/write and read/write conflicts are not specifically checked, but eliminated as a byproduct of the algorithm.

The following example illustrates instruction deferral and is a simplified version of Tomosulo's algorithm.

---

**Example 3.12**
Consider the operation $P = A * B + C * D$ where P, A, B, C, and D are memory operands. This can be executed by the sequence:

1. $r1 \leftarrow (A)$
2. $r2 \leftarrow (B)$
3. $r2 \leftarrow (r1) * (r2)$
4. $r3 \leftarrow (C)$
5. $r4 \leftarrow (D)$
6. $r4 \leftarrow (r3) * (r4)$
7. $r4 \leftarrow (r2) * (r4)$
8. $P \leftarrow (r4)$

Figure 3.26 shows the entries in the reservation stations for all the functional units. The tag fields indicate that the value for A is not

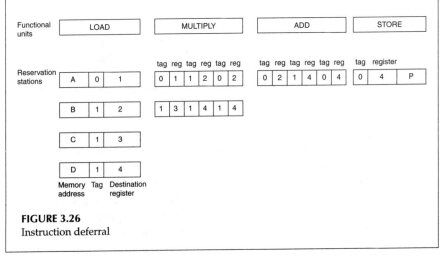

**FIGURE 3.26**
Instruction deferral

# Pipelining

> available in r1 at this time and hence operation 1 and hence 3, 7, and 8 are deferred until A arrives. Meanwhile the pipeline can proceed with other operations.

The pipeline design techniques, instruction deferral and forwarding mechanisms described in this chapter reflect the state of the art, although first implemented almost 20 years ago. But the inclusion of cache memory into computer systems has contributed to a drastic lowering of time required for memory access. It is now possible to have a large number of registers in the machine structure. With sophisticated compilers now available, it is possible to allocate this large number of registers judiciously thus eliminating a majority of read/write and write/write conflicts.

### 3.4.2 Performance Evaluation

In static pipelines, since all the initiations are of the same type, it is possible to design an optimal strategy for scheduling the pipeline to provide the maximum processing rate. In dynamic pipelines, where several RTs are used for initiation, several equally valid scheduling strategies are possible. Kogge (1981) identifies the following:

1. Maximizing the total number of initiations of any kind in any order per unit time.
2. Maximizing the total number of initiations per unit time given that out of this total a percentage must be from each type of RT and no particular order imposed on the sequence.
3. Minimizing the total time required to handle a specific sequence of initiation table types.

In an instruction processing pipeline of an SISD (single instruction stream, single data stream), the first criterion above corresponds to the maximum number of instructions per second that can be executed in any order; the second criterion imposes an instruction mix (i.e., arithmetic type 40%, Branches 10%, so on), and the third corresponds to the performance of some benchmark program. Thus, the criteria are increasingly restrictive.

Performance measurements of the first type are similar to those applied to static pipelines. The difference is that in computing the average latency of a cycle, the period is divided by the total number of entries in all initiation sets in the arcs making up the cycle. Also, since more than one reservation table may be used for initiations at any time, the lower bound for the average latency cannot be determined simply from the RTs. The following lemma provides the lower bound:

LEMMA: *The average latency of any cycle for a dynamic pipeline is bounded below by $1/N$, where $N$ is the maximum number of elements in any initiation set.*

The second performance measure requires the minimum average latency given a mix of initiations. Thomas and Davidson (1978) extend the solution for the static case to cover this measure. This method assumes that the scheduling algorithm is free to pick any of the RTs to make an initiation. Thus there is no restriction as to when a RT may be picked or in what order. This assumption reduces the complexity of analysis. The branch-and-bound and linear programming technique used in the analysis, computes for a given mix of initiations a collection of simple cycles that provide the MAL. But, the analysis does not guarantee that these cycles can be connected. That is, the controller should use additional paths to connect these cycles, thus making the control strategy less than optimal.

There is no known solution for the third performance measure, except for an exhaustive search process. A schedule that is optimal for one benchmark, may not be optimal for another bench mark. Usually variations of greedy strategy are used that at each stage chose from those arcs with appropriate initiation sets the one with the minimum latency.

## 3.5 Example Systems

Chapter 1 provided the details of pipeline operation in two modern day commercial architectures. A brief description of representative pipelined processors is given in this section. CDC STAR-100 is included for its historical interest. CDC-6600 is included to highlight its pipelined I/O system. CDC-6600 is also the basis for the Cray vector architecture described in Chapter 4. Sun Microsystem's Niagara microproccessor represents the latest breed of heavily pipelined architectures.

### 3.5.1 Control Data Corporation STAR-100

The CDC STAR-100 developed in late 1960s is a pipelined machine and served as the prototype for the Cyber series of machines manufactured by CDC. The structure of STAR-100 is shown in Figure 3.27. The CPU contains two multifunction arithmetic pipelines that can process 64-bit fixed- and floating-point operands and a string processing unit. Pipeline 1 performs addition and multiplication. Pipeline 2 performs addition, multiplication, division, and square-root extraction. It also contains a nonpipelined floating-point division unit. Each pipeline can function as a 64-bit pipeline or as two 32-bit pipelines working in parallel. Hence, there are in effect four 32-bit pipelines. Each of these can

*Pipelining* 115

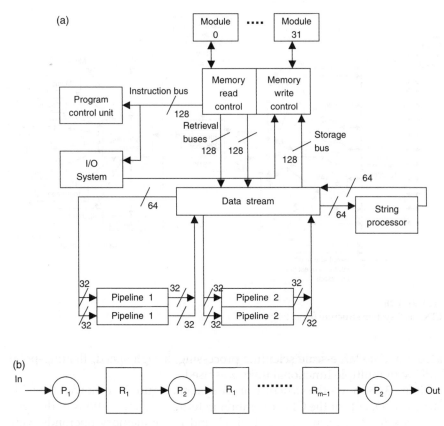

**FIGURE 3.27**
CDC STAR-100. (a) system structure and (b) pipeline

produce one floating-point output every 40 nanoseconds. Two 128-bit busses supply operands to the CPU from the memory and one 128-bit bus carries the results back to the memory. There is a fourth 128-bit bus is used for instruction fetch and I/O. Each of these busses can transmit 128 bits every 40 nsec.

The memory system consists of 32 modules, each with 2K words of 512 bits. The cycle time of the core memory modules is 1.28 $\mu$sec. The CPU pipelines produce a 128-bit result every 40 nsec or a total of 4K bits every 1.28 us. In order to sustain this rate, the memory has to output 2 streams of 4K bits and input 1 stream of 4K bits every cycle. In addition, another stream of 4K bits output is needed for instruction fetch, resulting in a total memory transfer rate of 16K bits every cycle. This rate is achieved by each of the 32 modules producing a 512-bit data every cycle.

### 3.5.2 Control Data Corporation 6600

The applications envisioned for CDC 6600 introduced in 1964 were: large-scale scientific processing and time-sharing of smaller problems. To

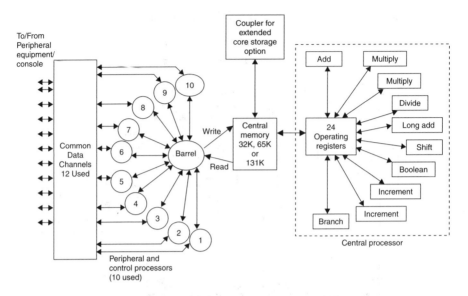

**FIGURE 3.28**
CDC 6600 system structure (Courtesy of CDC.)

accommodate large-scale scientific processing, a high-speed, floating-point CPU with multiple functional units was used.

Figure 3.28 shows the structure of the system. Figure 3.24 showed the CPU structure. Details of the CPU operation in terms of computations of the type $Z = A * B + C * D$, where $Z$, $A$, $B$, $C$, and $D$ are memory operands were given earlier in this chapter. The CPU obtains its programs and data from the central memory and can be interrupted by a peripheral processor. There are 24 operating registers in the CPU: eight 18-bit index registers, eight 18-bit address registers, and eight 60-bit floating-point registers interconnected by data and address paths. Instructions are either 15- or 30-bit long and there is an instruction stack that can hold up to 32 instructions to enhance the instruction execution speed.

The control unit maintains the scoreboard described earlier. The central memory is organized into 32 interleaved banks of 4K words each. Five memory trunks are provided between the memory and floating-point registers. An instruction calling for an address register implicitly initiates a memory reference on its trunk. An overlapped memory access and arithmetic operation is thus possible.

The I/O subsystem consists of twelve I/O channels controlled by 10 peripheral processors. One of the peripheral processors acts as the control unit of the system. The 10 peripheral processors access the central memory through the barrel mechanism, which has a 1000-nsec cycle. Each peripheral processor is connected to the memory for 100 nsec in each barrel cycle. During this 100-nsec interval the peripheral processor requests a memory access. The access is complete by the time the barrel visits that peripheral processor in the

# Pipelining

next cycle. The barrel is a ten-stage pipeline. The request from each peripheral processor enters this pipeline and gets processed in an overlapped manner over the next nine 100-nsec intervals. I/O channels are 12-bit bidirectional paths. One 12-bit word can be transferred in or out of the memory every 1000 nsec by each channel.

The major characteristics of the CDC-6600 that makes it versatile are: concurrent operation of the functional units within the CPU, high transfer rates between registers and memory, separation of CPU and I/O activities and a high bandwidth I/O structure. The Cray series of supercomputers also designed by Seymour Cray retain this basic architecture.

### 3.5.3 Sun Microsystem's Niagara Microprocessor (Geppert, 2005)

The Niagara to be released in 2006 is intended for the volume servers that are the heart of data centers running the information and web processing for businesses, universities, hospitals, factories, and the like. Niagara, named for the torrent of data and instructions that flow between the chip and its memory, was designed to do away with the impact of latency in data or instructions to arrive from memory.

Niagara uses a heavily pipelined architecture. Each stage of the pipeline performs one step of the instruction execution every clock cycle. In the first stage of the Niagara design, the pipeline gets an instruction from memory. The second stage selects that instruction for execution. The third stage determines what kind of instruction it is. The fourth stage executes the instruction. The fifth stage is used for getting data from memory. If the instruction does not access memory (but get their data from registers), it passes through the memory stage and on to the final stage, during which pipeline writes the results of the operation back into a register. Every clock cycle, a new instruction enters the pipeline. If all goes well, the instructions march through the pipeline in and one instruction is completed per clock cycle.

The performance of the processor is enhanced by two techniques: increasing the clock frequency and increasing the number of instructions the processor can execute in one-clock cycle. Clock frequencies have soared over the past 30-plus years by five orders of magnitude (from tens of kilohertz to more than 4 GHz). To increase the number of instructions per clock cycle, architects have added more pipelines. Some microprocessors such as Intel's Pentium 4, have eight pipelines, allowing it in principle to complete eight instructions in parallel during a single-clock cycle. Some designs have put the essential elements of two microprocessors on one piece of silicon. Such a dual-core microprocessor, with eight pipelines in each core running at 4 GHz, could execute 64 billion instructions per second if it could complete one instruction per pipeline per cycle. But, if an instruction in a pipeline needs data from memory, it has to wait until the data arrives before it can actually execute the instruction. The length of the individual delays depends on where the sought-after data is. If it is in the high-speed on-chip cache memory, the wait could be only a few clock cycles. But if the data is not in the cache, its retrieval from off-chip main memory may take hundreds or even thousands of clock cycles. Thus, minimizing this latency is perhaps the most important aspect of the processor design.

To improve throughput, two mechanisms have been used: executing some instructions in a different order from the way they occur in the instruction stream (out-of-order execution) or beginning the execution of instructions that may never be needed (speculative execution). Niagara uses the concept of *multithreading* to improve performance. It divides the instruction stream into several smaller streams, known as threads. The concept was first developed for Control Data Corp.'s CDC 6600 supercomputer the 1960s. In the Niagara design, each pipeline can handle four threads. In each cycle, the pipeline begins the execution of an instruction from a different thread. So, for example, when instruction from thread one is at stage 3 in the pipeline, an instruction from thread 2 will be at stage 2, and one from yet a different thread will be at stage 1. If the pipeline cannot continue to execute the thread-one instruction because it needs data from memory, it stores the information about the stalled instruction in a special type of on-chip memory called a register file. At the same time, it continues with the execution of the thread-two instruction, rotating among the three threads that are available. Then, when the needed data for the other thread becomes available, the pipeline jumps back to that thread, using the register file to pick up exactly where it left off.

In conventional microprocessors, architects obtain multigigahertz speeds by increasing the number of stages in the pipeline. Basically, a processing step that could be completed in one-clock cycle in a slower pipeline now needs two or more clock cycles to do the same job in a faster chip. But because of Niagara's ability to keep its pipelines running almost all of the time, the architects do not have to run the microprocessor at multigigahertz speeds in order to get good performance. And the slower speeds translate to a simpler pipeline. The simpler pipeline and lower frequency let Niagara run at much lower power than comparable microprocessors.

Niagara uses the Sun Sparc architecture. The approach was to build a very simple, straightforward pipeline, maximize the number of threads the pipeline can efficiently handle, put it into a microprocessor core, and then maximize the number of cores on the chip. Each Niagara chip has eight cores, which use the Sparc instruction set, and the pipeline in each core can switch among four threads. So a single-Niagara chip can handle 32 threads and execute eight instructions per clock cycle. The chip rarely wastes time waiting for data to come back from memory.

To further reduce latency, Niagara's architects have used a fat, fast interface between the processor cores and its two different kinds of on-chip memory. Each chip has two levels of memory. Every core has its own level-one memory, and the whole group of cores shares a larger, level-two memory. The level-two memory can send data to the cores at an aggregate rate of gigabytes per second.

## 3.6  Summary

Pipelining techniques have been adopted extensively to enhance the performance of serial processing structures. This chapter covers the basic principles and design methodologies for pipelines. Almost all the pipeline systems available today utilize static pipelines. A brief introduction to dynamic pipelines, which are more complex to design, is also given in this chapter. Although the throughput of the machine is improved by employing pipeline techniques, the basic cycle time of instructions remain as in nonpipelined implementations. In earlier machines with little or no pipelining, the average clock per instruction (CPI) was 5 to 10. Modern RISC architectures have achieved a CPI value close to 1. Pipelines exploit instruction-level parallelism in the program, whereby instructions that are not dependent on each other are executed in an overlapped manner. If the instruction-level parallelism is sufficient to keep the pipeline flowing full (with independent instruction) we achieve the ideal machine with a CPI of 1.

Three approaches have been tried to improve the performance beyond the ideal CPI case: *superpipelining*, *superscalar*, and *very long instruction word (VLIW)* architectures (Jouppi, 1989). A *superpipeline* uses deeper pipelines and control mechanisms to keep many stages of the pipeline busy concurrently. Note that as the latency of a stage gets longer, the instruction issue rate will drop and also there is a higher potential for data interlocks. A *superscalar* machine allows the issue of more than one instruction per clock into the pipeline, thus achieving an instruction rate higher than the clock rate. In these architectures, the hardware evaluates the dependency among instructions and dynamically schedules a packet of independent instruction for issue at each cycle. In a *VLIW* architecture, a machine instruction corresponds to a packet of independent instructions, created by the compiler. No dynamic hardware scheduling is used.

Pipeline techniques are extensively used in all architectures today, from microprocessors to supercomputers. This chapter provided some details of two historic pipelined machines, along with one modern architecture. Chapter 1 covered the pipelining details of two microprocessors. Chapter 4 expands on pipeline techniques as used in building vector processors.

## Problems

3.1. Describe the conditions under which an $n$-stage pipeline is $n$ times faster than a serial machine.

3.2. Design a pipeline ALU that can add two 8-bit integers. Assume that each stage can add two 2-bit integers and employ sufficient number of stages. Show the operation of the pipeline using an RT.

3.3. Design a pipeline for multiplying two floating-point numbers represented in IEEE standard format. Assume that addition requires $T$ sec and multiplication requires $2T$ sec and shift takes $T/2$ sec. Use as many stages as needed and distribute the functions such that each stage approximately consumes the same amount of time.

3.4. Compute the minimum number of pipeline cycles required to multiply 50 floating-point numbers in the pipeline of Problem 3.3, assuming that the output from the last stage can be fed back to any of the stages that requires it, with the appropriate delay.

3.5. For the following RT, derive the CV and identify the maximum rate cycle:

|    | 1 | 2 | 3 | 4 | 5 | 6 | 7 |
|----|---|---|---|---|---|---|---|
| S1 | X | X |   | X |   |   |   |
| S2 |   | X |   |   |   |   | X |
| S3 |   |   | X |   |   |   |   |
| S4 |   |   |   |   | X |   |   |
| S5 |   |   |   |   |   | X |   |

3.6. For each of the following CVs, draw a reduced state diagram and determine the maximum performance possible:

1. 110010100
2. 100101
3. 10101000

3.7. For any computer system you are familiar with, analyze the instruction cycle and split that into a three-stage pipeline in which the stages perform fetch, address compute, and execute functions. What types of problems you would encounter?

3.8. It is required to multiply two ($n \times n$) matrices. Design a pipeline with a one-stage adder and a two-stage multiplier where each stage consumes $T$ sec and compute the execution time for matrix multiplication.

3.9. The following program adds 2 to each element of a vector located starting at memory location $V$. The index register $X$ contains the number of elements in the vector:

```
LOOP  LDA  V(X)      Load an element into the Accumulator
      ADD  =2        Add 2 to Accumulator
      STA  V(X)      Store it back
      DEX            Decrement Index register
      BNZ  LOOP      Branch back if X is not 0
```

Show the execution characteristics of this code on the pipeline of Figure 3.2. Schedule instructions appropriately to enhance the performance of the pipeline.

3.10. One common mechanism used to enhance pipeline performance while executing loops is called *loop unrolling*. That is, the body of the loop is repeated to form a program with strict sequential instruction stream. Assume that the number of elements in Problem 3.10 is 8:

1. Unroll the loop such that two elements are processed each time through the loop. Determine the performance of the pipeline with appropriate instruction scheduling.
2. Unroll the loop completely. That is, no looping at all. Examine the performance of the pipeline, with appropriate scheduling.

3.11. Assume that a CPU uses four cycles (fetch, calculate address, memory read, move the data to the destination register) for a LOAD register instruction; correspondingly four cycles for a STORE instruction and three cycles (fetch, compute results, move them to the destination register) for other register to register instructions. Show the reservation pattern for the following program:

```
LOAD    Z       ACC ← Z
ADD     R1      ACC ← ACC + R1
STORE   Z       Z   ← ACC
SUB     R2      ACC ← ACC - R2
```

where, ACC, R1, and R2 are registers and Z is a memory location. How many cycles are needed to execute this program, assuming maximum instruction overlap?

3.12. Assume that an $n$-element vector operation takes $(s \times n)$ cycles on a nonpipelined machine and $(f + p \times n)$ cycles on a pipelined machine, where $f$ is the pipeline fill time and $p$ is the time taken by stages in the pipeline (in cycles). For what values of $n$, the pipelined implementation is faster?

3.13. In implementing the delayed branch technique, the compiler must determine when an instruction can be moved into the branch delay slot. What conditions are appropriate for the compiler to consider?

3.14. Compare the relative merits of "superpipelined" and "superscalar" architectures. To what application environment each technique is suitable? What is an "underpipelined" architecture?

# References

Geppert, L. Sun's big splash. *IEEE Spectrum*, 42, 2005, 56–60.

Handler, W. The impact of classification schemes on computer architecture. In *Proceedings of the International Conference on Parallel Processing*, pp. 7–15, 1977.

Jouppi, N. P. and Wall, D. W. Available instruction-level parallelism for superscalar and superpipelined machines. In *Proceedings of the IEEE/ACM Conference on Architectural Support for Programming Languages and Operating Systems*, pp. 272–282, April 1989.

Kogge, P. M. *The Architecture of Pipelined Computers*. New York: McGraw-Hill, 1981.

Losq, J. J., Rao, G. S., and Sachar, H. E. Decode history table for conditional branch instructions. *U.S. Patent No. 4,477,872*, October 1984.

Patel, J. H. and Davidson, E. S. Improving the throughput of a pipeline by insertion of delays. In *Proceedings of the Third Annual Computer Architecture Symposium*, pp. 159–163, 1976.

Ramamoorthy, C. V. and Li, H. F. Pipeline architectures. *ACM Computing Surveys*, 9, 1977, 61–102.

Shar, L. E. Design and scheduling of statistically configured pipelines. *Digital Systems Lab Report* SU-SEL-72-042, Stanford University, Stanford, CA, September 1972.

Smith, J. E. and Plezkun, A. R. Implementing precise interrupts in pipelined processors. *IEEE Transactions on Computers*, 37, 1988, 562–573.

Stone, H. S. *High Performance Computer Architecture*. Reading, MA: Addison-Wesley, 1987.

Sussenguth, E., Instruction Sequence Control. *U.S. Patent No. 3,559,183*, January 26, 1971.

Thomas, A. T. and Davidson, E. S. Scheduling of multiconfigurable pipelines. In *Proceedings of the 12th Annual Allerton Conference on Circuits and System Theory*, pp. 658–669, 1978.

Thornton, J. E. *Design of a Computer: The Control Data 6600*. Glenview, IL: Scott, Foresman, 1970.

# 4

## Vector Processors

Manipulation of arrays or vectors is a common operation in scientific and engineering applications. Typical operations on array-oriented data are:

1. Processing one or more vectors to produce a scalar result (as in, computing the sum of all the elements of the vector, finding the maximum or minimum element, etc.).
2. Combining two vectors to produce a third vector (as in, addition or multiplication of two arrays element-by-element).
3. Combining a scalar and a vector resulting in a vector (as in, multiplying each element of the vector by a scalar).
4. A combination of the above three operations.

Two architectures suitable for the vector processing environment have evolved over the years: pipelined vector processors and parallel array processors.

*Pipelined vector processors* (or pipelined array processors, or simply "vector processors") covered in this chapter, utilize one or more pipelined processors to achieve high computation throughput. Examples of this type of architecture are the supercomputers from Cray (Cray-1, Cray-2, Cray X-MP, Cray Y-MP, Cray-3, Cray-4, Cray X1), Hitachi (Super Technical Servers), and NEC (SX series).

*Parallel array processors* covered in Chapter 5, adopt a multiplicity of processors that operate on elements of arrays in parallel, in an instruction lock-step mode. This is the class of SIMD architectures presented in Chapter 2. These architectures exploit the *data-parallelism* in array-oriented computations.

Vector processors are supercomputers optimized for fast execution of long groups of vectorizable scientific code, operating on large datasets. Since the highest possible computation rate is the main criteria behind building these machines and not the cost optimization (in general), these are often tagged "cost-is-no-object" approach to computer architecture. Features that are possible only on these expensive systems, eventually trickle down to commodity microprocessors, as the hardware technology progresses. As such, these machines could be good indicators of the types of things to come, in future processors. Vector processors are extensively pipelined architectures designed to operate on array-oriented data. In these processors, the operations on the arrays take place serially on each element of the array, rather than all the elements in parallel. The individual elements are streamed into the extensively pipelined CPU (Central Processing Unit), to operate upon. The memory system is also deeply pipelined and interleaved, to meet the demand of the pipelined CPU. In addition, the register set of the CPU may be large enough to hold all the elements of one or more vectors. This combination of extensively pipelined CPU and memory, and large register set, provides the high computation throughput achieved by these architectures.

The models for two common vector processor architectures are provided in the next section, followed by memory design considerations, an important aspect of vector processor design, in Section 4.2. Section 4.3 utilizes the Cray series architecture to illustrate the features of practical architectures. Performance evaluation and programming concepts are covered in Sections 4.4 and 4.5, respectively. Brief descriptions of three commercial architectures are provided in Section 4.6.

## 4.1 Vector Processor Models

The models for two common vector processor architectures are introduced in this section. The following examples illustrate the advantages of vector instruction coding.

---

**Example 4.1**

Consider element-by-element addition of two $N$-element vectors **A** and **B** to create the sum vector **C**. That is:

$$\mathbf{C}_i = \mathbf{A}_i + \mathbf{B}_i \quad 1 \leq i \leq N \tag{4.1}$$

This computation can be implemented on an SISD by the following program:

```
For i = 1, N
 C[i] = A[i] + B[i]
endfor
```

Assuming two machine instructions for loop control and four machine instructions to implement the assignment statement (Read A, Read B, Add, Write C), the execution time of this program is $(6 \times N \times T)$ where $T$ is the average instruction cycle time of the machine.

# Vector Processors

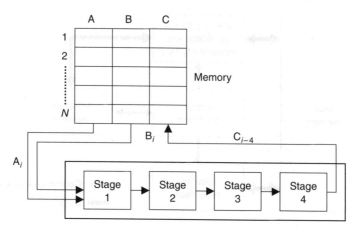

**FIGURE 4.1**
Vector computational model

Figure 4.1 shows a pipelined implementation of this computation using a four-stage addition pipeline with a cycle time $T$. Now the program consists of only one instruction:

$$C[i] = A[i] + B[i] \quad (1 \leq i \leq N)$$

The loop overhead of the SISD program above is thus eliminated and only one instruction needs to be fetched from the memory. The total execution time is the sum of one instruction fetch and decode time (assumed to be $T$) and the time to add $N$ pairs of operands using the pipeline.

The performance of a pipelined functional unit is characterized by the *start-up time*, which is the number of clock cycles required prior to the generation of the first result or simply the depth of the pipeline in clock cycles. After the start-up time the functional unit can deliver one result every clock cycle. Thus, the time to complete an $N$-element vector operation in a pipeline is:

$$\text{Start-up time} + (N-1) * \text{Initiation rate} \quad (4.2)$$

The *start-up time* in the above pipeline is $4T$ and one operation can be initiated every cycle. Thus the above vector operation can be completed in $(4 + N - 1) * T$ cycles. The total execution time thus is:

$$= T + (4 + N - 1) * T$$
$$= (4 + N)T$$

Hence, the speedup due to pipelined implementation is:

$$\frac{6NT}{(4+N)T} \quad (4.3)$$

This implies a sixfold speedup for large values of $N$.

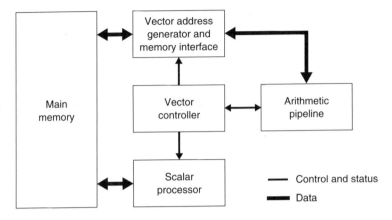

**FIGURE 4.2**
Memory-oriented vector processor

Note from (4.2) above, that the start-up time adds a considerable overhead for small values of $N$. For large values of $N$, the effect of start-up time is negligible and the *completion rate* (i.e., the rate at which the results are produced) tends toward the initiation rate.

The pipelined array processor structure of Figure 4.1 can be generalized to that shown in Figure 4.2 to represent the architecture of contemporary vector processors. The main memory holds instructions and data. Its organization is a key factor in the performance of the system, since vector processors retrieve and store large amounts of data from and to the memory. The memory is usually very heavily interleaved to accommodate the fast access requirements of vector data. Note for instance that the pipeline organization in Figure 4.1 requires two read and one write operation from the memory at each cycle. Memory system design should accommodate this access requirement for data fetch and store operations in addition to instruction fetch access requirements.

The scalar processor handles all scalar aspects of processing such as program control and vector setup, operating system functions and input/output (I/O). It calls upon the vector controller to handle vector computations.

The vector controller decodes vector instructions, computes operand address parameters, and initiates the vector memory controller and the arithmetic pipeline operations and monitors them. At the end of the vector instruction execution, it performs any cleanup and status gathering needed.

The function of the vector address generator and memory interface module is to provide a high-speed transfer of data between the arithmetic pipeline and the memory. It receives the operand address parameters from the vector controller and converts them into a series of memory access patterns that access the memory as fast as possible.

The arithmetic pipeline receives the operands from the memory and returns the results to the memory. In practice, there may be more than one arithmetic pipeline in the system. These pipelines could each be dedicated to a single function or may each handle multiple functions.

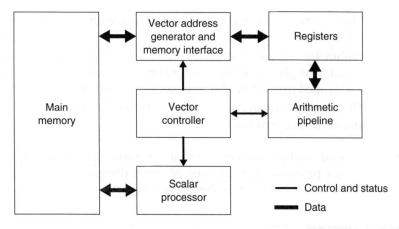

**FIGURE 4.3**
Register-oriented vector processor

The above model represents the *memory-oriented* vector processor architecture. Figure 4.3 shows the *register-oriented* vector processor architecture commonly used to build modern day supercomputers. In this architecture, scalar and vector data are first brought into a set of registers and the arithmetic pipeline receives the operands from these registers and stores the results into them. Since accessing registers is faster than accessing memory, pipeline throughput would be enhanced, compared with the structure of Figure 4.2. Typically, vector operands are transferred between the main memory and the registers in blocks and the arithmetic pipeline accesses these operands from the registers element by element. Thus, the registers serve as buffers between the memory and the pipeline. This buffering also allows the simultaneous operation of the pipeline–register and register–memory interfaces. The Cray and Hitachi machines are examples of register-oriented architecture.

The following characteristics of vector processor architectures contribute to their high performance:

1. High-speed memory that allows retrieval and storage of voluminous data.
2. A large number of processor registers with vector registers capable of holding all the elements of vectors of typical length and scalar registers for non-vector data and addresses.
3. Instruction set containing vector-oriented instructions that define operations on the entire vector or vectors.
4. A multiplicity of overlapped processing levels:
   (a) Each arithmetic pipeline provides overlapped operations.
   (b) If multiple pipelines are employed, they can all be operating in parallel.

(c) The scalar processor overlaps its operations with those of the vector module.

(d) Within the vector module, the vector controller and the vector address generator might be performing their functions in an overlapped manner on different vector instructions.

(e) The memory–register and register–pipeline interfaces operate in an overlapped manner.

As mentioned earlier, memory system performance is an important aspect of vector processor architectures. Memory design considerations are described next. Chapter 5 extends these concepts further.

## 4.2 Memory Design Considerations

Memory *bandwidth* depends on the memory system configuration, memory module configuration, and the processor architecture. Memory system configuration is characterized by the number of memory modules, bus width and the address decoding structure (i.e., interleaved, banked, wider-word fetch, etc.). The module characteristics include the size, access time, and cycle time. Memory bandwidth must match the demand of multiple pipelined processors, if the vector processor has to deliver maximum performance.

---

**Example 4.2**
Consider a vector processor with four 32-bit floating-point processors, each requiring two 32-bit operands and produce one 32-bit result every clock cycle. Assume that one 32-bit instruction is fetched for each arithmetic operation. The memory thus needs to support reading $(2*4+1) \times 32 = 288$ bits and writing $4*32 = 128$ bits every cycle, resulting in a total traffic of 416 bits/cycle. Because each set of operands consists of four 32-bit elements, the bus width to main memory for each access will be $4*32 = 128$ bits. Also, since there are two operand fetches, one instruction fetch and one result storage every cycle, the main memory should have four, 128-bit wide unidirectional buses. If the memory cycle time is 1.28 $\mu$sec and the processor cycle time is 40 nsec, we would need 1.28 $\mu$sec/40 nsec or 32 memory modules to match the demand rate.

---

Commercial vector processors have adopted two mechanisms to meet the above demand requirements: memory systems configured with multiple memory modules allowing simultaneous access and inserting fast (register and cache) intermediate memories between the pipelines and the main memory.

# Vector Processors

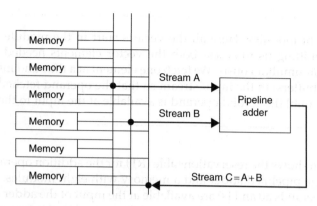

**FIGURE 4.4**
Vector processor with multiple three-port memory modules

Consider the memory structure shown in Figure 4.4. Here the main memory is considered to be composed of multiple modules with each module having three ports corresponding to the three data streams associated with the pipeline. All the streams can be active simultaneously, but only one port of a module can be used at a time.

**Example 4.3**
Consider the vector addition of equation (4.1). Figure 4.5 shows the data structure in a memory system (of the type shown in Figure 4.4)

| Module 0 | a0 | b0 | c0 | a8 | b8 | c8 | ........ |
| --- | --- | --- | --- | --- | --- | --- | --- |
| 1 | a1 | b1 | c1 | a9 | b9 | c9 | ........ |
| 2 | a2 | b2 | c2 | a10 | b10 | c10 | ........ |
| 3 | a3 | b3 | c3 | a11 | b11 | c11 | ........ |
| 4 | a4 | b4 | c4 | a12 | b12 | c12 | ........ |
| 5 | a5 | b5 | c5 | a13 | b13 | c13 | ........ |
| 6 | a6 | b6 | c6 | a14 | b14 | c13 | ........ |
| 7 | a7 | b7 | c7 | a15 | b15 | c15 | ........ |

**FIGURE 4.5**
Data structure for the computation of $C = A + B$

with eight modules. Here all the vectors start in the module 0, thus not permitting us to access both the vector elements needed for the operation simultaneously. We fetch one element at a time and introduce delays (buffers) in the input stream so that the operand fetched first is delayed until the second operand is available at the input to the adder.

Figure 4.6 shows the reservation table (RT) for the addition operation using a three-stage pipelined adder and a memory with eight modules. At time 2, both the operands a0 and b0 are available at the input of the adder and hence the result of addition is available at time 5. Since module 0 is free at this time, the result is written into it. Then on, one result is written into the successive modules of the memory every cycle. In this figure the operands fetched are shown by the corresponding element names (a0, a1, b0, b1, etc.) and the operation within the pipeline stages is shown by its index value (0, 1, 2, etc.) and the corresponding results are denoted as (w0, w1, etc.). As can be seen, this organization of the memory system and data structure requires the insertion of one delay at the A input of the adder.

|    | 0  | 1  | 2  | 3  | 4  | 5  | 6  | 7  | 8  | 9  | 10 |
|----|----|----|----|----|----|----|----|----|----|----|----|
| M0 | a0 | b0 |    |    |    | w0 |    |    | a8 | b8 |    |
| M1 |    | a1 | b1 |    |    |    | w1 |    |    | a9 | b9 |
| M2 |    |    | a2 | b2 |    |    |    | w2 |    |    |    |
| M3 |    |    |    | a3 | b3 |    |    |    | w3 |    |    |
| M4 |    |    |    |    | a4 | b4 |    |    |    | w4 |    |
| M5 |    |    |    |    |    | a5 | b5 |    |    |    | w5 |
| M6 |    |    |    |    |    |    | a6 | b6 |    |    |    |
| M7 |    |    |    |    |    |    |    | a7 | b7 |    |    |
| S1 |    |    |    | 0  | 1  | 2  | 3  | 4  | 5  |    |    |
| S2 |    |    |    |    | 0  | 1  | 2  | 3  | 4  | 5  |    |
| S3 |    |    |    |    |    | 0  | 1  | 2  | 3  | 4  | 5  |

8 Modules; 1-cycle Read/Write;
3 stage pipeline; 1 delay on A;
no conflict on the output

**FIGURE 4.6**
Reservation table (three-stage adder)

# Vector Processors

**Example 4.4**

Now consider the RT of Figure 4.7 for an architecture with a six-module memory, three-stage pipelined adder, and a memory access time equivalent to two processor cycle times. All arrays start in memory module 0 as before. Because each memory access takes two cycles, the first operand from module 0 will be available at the end of clock cycle 1. We can initiate the fetch of the operand b0 from module 0 at this time and the operand will be available at the end of clock cycle 3. In the meantime we need to buffer the operand a0 for 2 clock cycles. This is done by introducing delay elements in the input stream A. The result of this operand pair is available at time 7, but cannot be written into module 0 until time 10, since module 0 is busy fetching operands at times 7 through 9. Thus, another three unit delay is needed at the output of the adder.

|    | 0  | 1  | 2  | 3  | 4  | 5  | 6  | 7  | 8  | 9  | 10 | 11 | 12 | 13 | 14 |
|----|----|----|----|----|----|----|----|----|----|----|----|----|----|----|----|
| M0 | a0 | a0 | b0 | b0 |    |    | a6 | a6 | b6 | b6 | w0 | w0 |    |    |    |
| M1 |    | a1 | a1 | b1 | b1 |    |    | a7 | a7 | b7 | b7 | w1 | w1 |    |    |
| M2 |    |    | a2 | a2 | b2 | b2 |    |    |    |    |    |    | w2 | w2 |    |
| M3 |    |    |    | a3 | a3 | b3 | b3 |    |    |    |    |    | w3 | w3 |    |
| M4 |    |    |    |    | a4 | a4 | b4 | b4 |    |    |    |    |    |    |    |
| M5 |    |    |    |    |    | a5 | a5 | b5 | b5 |    |    |    |    |    |    |
| S1 |    |    |    |    | 0  | 1  | 2  | 3  |    |    |    |    |    |    |    |
| S2 |    |    |    |    |    | 0  | 1  | 2  | 3  |    |    |    |    |    |    |
| S3 |    |    |    |    |    |    | 0  | 1  | 2  | 3  |    |    |    |    |    |

6 Modules; 3 delays in A; 2-cycle Read/Write
3-stage pipeline; 3 delays on the output

**FIGURE 4.7**
Reservation table

Figure 4.8 shows the general structure of the vector processor with delay elements inserted in the input and the output streams.

It is also possible to enhance the memory bandwidth by structuring the data such that the data elements required at any time are in separate physical modules. For instance, array $A$ in the above example could be stored starting in module 0 and array $B$ starting in module 1. Then both a0 and b0 could be accessed simultaneously. But, based on the pipeline characteristics, the results must be stored in a module that is free when they are available. Such storage pattern may not in general, result in a data structure that allows conflict free access to the resultant data for subsequent operations.

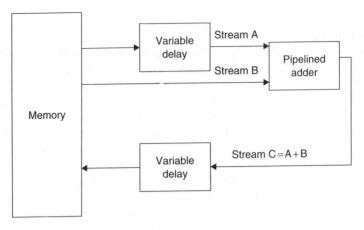

**FIGURE 4.8**
Vector processor with variable delays at the input and output data streams

The above examples illustrate the importance of structuring data in the memory to reduce memory bank conflicts. Chapter 5 discusses this topic further.

Assuming that data is structured in the memory with $N$ banks in an interleaved fashion (i.e., consecutive data element lie in different physical banks of the memory), one efficient way to fetch $N$ elements is to read all the $N$ banks simultaneously, thus fetching $N$ elements in one cycle. To transfer these $N$ elements simultaneously to the processor, an $N$-wordwide bus would be needed. One common way to reduce the bus width is to read the memory in a phased manner. That is, the $N$ words are read simultaneously during the first read operation. The data words are transferred to the processor one at a time. As soon as a word is transferred, a read is initiated in the corresponding memory bank. The phased read operation between the banks of the memory is continued until all the data needed are transferred.

A common method of increasing the memory system bandwidth further is to insert high-speed intermediate memory between the main memory and the processor pipelines. The intermediate memory can be organized as either a cache or as a set of individually addressable registers. The register organization is preferable because of the overhead imposed by the cache search mechanism. The register organization also allows the accessing of individual elements of a vector directly, since each element would lie in an individually addressable register. If the vector dataset required for a computation is not stored in the memory properly, memory references would be scattered (i.e., poor "locality"), thus making the cache organization inefficient. Also, a vector instruction could reference multiple data elements simultaneously. Searching the cache for multiple elements results in a high overhead.

The Cray series of vector processors described in the next section utilize multiple registers as intermediate memory for data access and cache mechanism for instruction fetching.

# Vector Processors

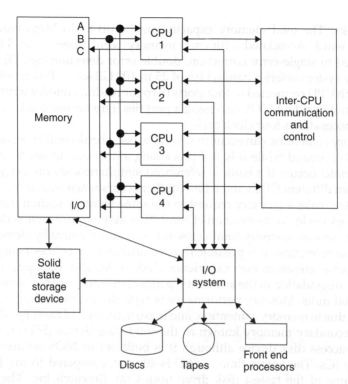

**FIGURE 4.9**
Cray X-Mp/4 structure

## 4.3 Architecture of the Cray Series

The CRAY-1, a second-generation vector processor from Cray Research Inc., has been described as the most powerful computer of the late 1970s. Benchmark studies show that it is capable of sustaining computational rates of 138 MFLOPS over long periods of time and attaining speeds of up to 250 MFLOPS in short bursts. This performance is about 5 times that of the CDC 7600 or 15 times that of an IBM System/370 Model 168. Figure 4.9 shows the structure of the Cray X-MP (successor to Cray-1). A four-processor system (X-MP/4) is shown. The Cray X-MP consists of four sections: multiple Cray-1-like CPUs, the memory system, the input/output (I/O) system, and the processor interconnection. The following paragraphs provide a brief description of each section.

### 4.3.1 Memory

The memory system is built out of several sections, each divided into banks. Addressing is interleaved across the sections and within sections across the banks. The total memory capacity can be up to 16 Megawords with 64 bits/word. Associated with each memory word, there is an 8-bit field dedicated to single-error correction/double-error detection (SECDED). The memory system offers a bandwidth of 25 to 100 GB sec$^{-1}$. It is multiported, with each CPU connected to four ports (two for reading, one for writing, and one for independent I/O). Accessing a port ties it up for one clock cycle, and a bank access takes four clock cycles.

Memory contention can occur in several ways: a bank conflict occurs when a bank is accessed while it is still processing a previous access; a simultaneous conflict occurs if a bank is referenced simultaneously on independent lines from different CPUs and a line conflict occurs when two or more of the data paths make a memory request to the same memory section during the same clock cycle. Memory conflict resolution may require wait states to be inserted. Because memory conflict resolution occurs element by element during vector references, it is possible that the arithmetic pipelines being fed by these vector references may experience clock cycles with no input. This can produce degradation in the arithmetic performance attained by the pipelined functional units. Memory performance is typically degraded by 3 to 7% on average due to memory contention, and in particularly bad cases by 20 to 33%.

The secondary memory known as the solid-state device (SSD) is used as a faster-access disk device although it is built out of MOS random access memory ICs. The access time of SSD is 40 $\mu$sec, compared to the 16 msec access time of the fastest disk drive from Cray Research Inc. The SSD is used for the storage of large-scale scientific programs that would otherwise exceed main memory capacity and to reduce bottlenecks in I/O-bound applications. The central memory is connected to the SSD through either one or two 1000 MB sec$^{-1}$ channels. The I/O subsystem is directly connected to the SSD thereby allowing prefetching of large datasets from the disk system to the faster SSD.

### 4.3.2 Processor Interconnection

The interconnection of the CPUs assumes a coarse-grained multiprocessing environment. That is, each processor (ideally) executes a task almost independently, requiring communication with other processors once every few million or billion instructions. Processor interconnection is comprised of the clustered share registers. The processor may access any cluster that has been allocated to it in either user or system monitor mode. A processor in monitor mode has the ability to interrupt any other processor and cause it to go into monitor mode.

### 4.3.3 Central Processor

Each CPU is composed of low-density Emitter Coupled Logic (ECL) with 16 gates/chip. Wire lengths are minimized to cut the propagation delay of

# Vector Processors

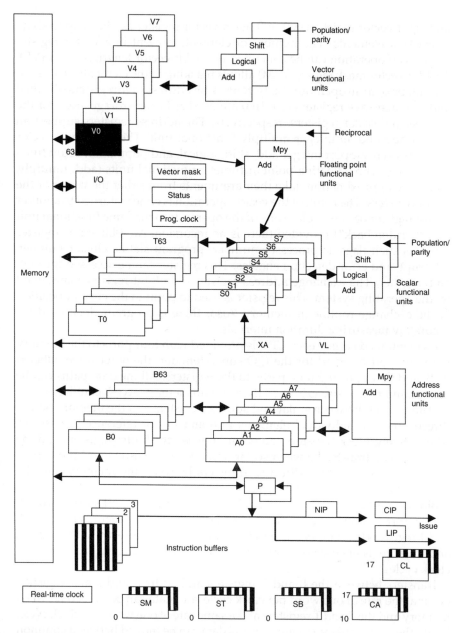

**FIGURE 4.10**
Cray X-MP CPU (Courtesy of Cray Research Inc.)

signals to about 650 psec. Each CPU is a register-oriented vector processor (Figure 4.10) with various sets of registers that supply arguments to and receive results from several pipelined, independent functional units. There are eight 24-bit address registers (A0–A7), eight 64-bit scalar registers (S0–S7), and eight vector registers (V0–V7). Each vector register can hold up to sixty-four 64-bit elements. The number of elements present in a vector register for a given operation can be contained in a 7-bit vector length register (VL). A 64-bit vector mask register (VM) allows masking of the elements of a vector register prior to an operation. Sixty-four 24-bit address save registers (B0–B63) and 64 scalar save registers (T0–T63) are used as buffer storage areas for the address and scalar registers, respectively. The address registers support an integer add and an integer multiply functional unit. The scalar and vector registers each support integer add, shift, logical, and population count functional units. Three floating point arithmetic functional units (add, multiply, reciprocal approximation) take their arguments from either the vector or the scalar registers. The result of the vector operation is either returned to another vector register or may replace one of the operands to the same functional unit (i.e., "written back") provided there is no recursion. An 8-bit status register contains such flags as processor number, program status, cluster number, interrupt, and error detection enables. This register can be accessed through an S register. The exchange address register is an 8-bit register maintained by the operating system. This register is used to point to the current position of the exchange routine in memory. Also, there is program clock used for accurately measuring duration intervals.

As mentioned earlier, each CPU is provided with four ports to the memory with one port reserved for the I/O subsystem and the other three, labeled A, B, and C supporting data paths to the registers. All the data paths can be active simultaneously, as long as there are no memory access conflicts.

Data transfer between scalar and address registers and the memory occurs directly (i.e., as individual elements into and out of referenced registers). Alternatively, block transfers can occur between the buffer registers and the memory. The transfer between scalar and address registers and the corresponding buffers is done directly. Transfers between the memory and the vector registers are done only directly.

Block transfer instructions are available for loading to and storing from B (using port A) and T (using port B) buffer registers. Block stores from the B and T registers to memory use port C. Loads and stores directly to the address and scalar registers use port C at a maximum data rate of one word every two clock cycles.

Transfers between the B and T registers and address and scalar registers occur at the rate of one word per clock cycle, and data can be moved between memory and the B and T registers at the same rate of one word per clock cycle. Using the three separate memory ports data can be moved between common memory and the buffer registers at a combined rate of three words per clock cycle, one word into B and T and one word from one of them.

The functional units are fully segmented (i.e., pipelined), which means that a new set of arguments may enter a unit every clock period (8.5 nsec). The segmentation is performed by holding the operands entering the unit and the partial results at the end of every segment until a flag allowing them to proceed is set. The number of segments in a unit determines

**TABLE 4.1**

Functional unit characteristics (Cray X-MP)

| Type | Operation | Registers used | Number of bits | Unit time (clock periods) |
|---|---|---|---|---|
| Address | Integer add | A | 24 | 2 |
|  | Integer multiply | A | 24 | 4 |
| Scalar | Integer add | S | 64 | 3 |
|  | Shift | S | 64 | 2 or 3 |
|  | Logical | S | 64 | 1 |
|  | Population | S | 64 | 3 or 4 |
|  | Parity and leading zero | S&A | 64 | 3 or 4 |
| Vector | Integer add | V | 64 | 3 |
|  | Shift | V&A | 64 | 3 or 4 |
|  | Logical | V | 64 | 2 |
|  | Second logical | V | 64 | 3 |
|  | Population and parity | V | 64 | 6 |
| Floating point | Add | S or V | 64 | 6 |
|  | Multiply | S or V | 64 | 7 |
|  | Reciprocal | S or V | 64 | 14 |
| Memory | Scalar load | S | 64 | 14 |
| Transfer | Vector load (64-element) | V | 64 | 17 + 64 |

the start-up time for that unit. Table 4.1 shows the functional unit characteristics.

**Example 4.5**

Consider again the vector addition:

$$C[i] = A[i] + B[i] \quad 1 \leq i \leq N$$

Assume that $N$ is 64, that $A$ and $B$ are loaded into two vector registers (V0, V1), and the result vector is stored in another vector register (V2) as shown in Figure 4.11(a). The unit time for floating-point addition is six clock periods. Including one clock period for transferring data from vector registers to the add unit and one clock to store the result into another vector register as shown in Figure 4.11(b), it would take $(64 \times 8 = 512)$ clock periods to execute this addition in scalar mode.

The first element of the result will be stored into the vector register after eight clock periods. Afterwards there will be one result every clock period. Therefore the total execution time is $(8 + 63 = 71)$ clock periods.

If $N < 64$, the above execution times would be reduced correspondingly. If $N > 64$, the computation is performed on units of 64 elements at a time. For example, if $N$ is 300, the computation is performed on four sets of 64, elements each, followed by the final set with the remaining 44 elements.

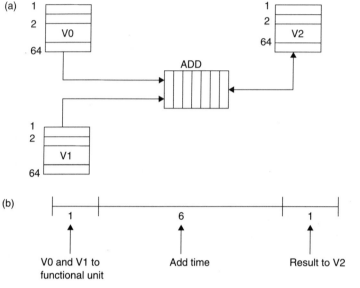

**FIGURE 4.11**
Vector pipelining on Cray X-MP: (a) pipeline and (b) timing

The vector length register contains the number of elements ($n$) to be operated upon at each computation. If $M$ is the length of vector registers in the machine, the following program can be used to execute the above vector operation for an arbitrary value of $N$.

```
begin = 1
n = (N mod M)
for i = 0, (N/M)
   for j = begin, begin + n-1
       C[j] = A[j] + B[j]
   endfor
begin = begin + n
n = M
endfor
```

Here, first the ($N$ mod $M$) elements are operated upon, followed by $N/M$ sets of $M$ elements.

In practice, the vector length will not be known at compile time. The compiler generates the code similar to above, such that the vector operation is performed in sets of length less than or equal to $M$. This is known as

# Vector Processors

*strip mining*. Strip mining overhead must also be included in start-up time computations of the pipeline.

In order for multiple functional units to be active simultaneously, intermediate results must be stored in the CPU registers. When properly programmed, the Cray architecture can arrange CPU registers such that the results of one functional unit can be input to another independent functional unit. Thus, in addition to the pipelining within the functional units, it is possible to pipeline arithmetic operations between the functional units. This is called *chaining*.

Chaining of vector functional units and their overlapped, concurrent operation are important characteristics of this architecture that bring about a vast speed-up in the execution times. The following example shows a loop where overlapping would occur.

---

**Example 4.6**
Consider the loop:

```
For J = 1,64
  C(J) = A(J) + B(J)
  D(J) = E(J) * F(J)
Endfor
```

Here, the addition and multiplication can be done in parallel because the functional units are totally independent.

---

The following example illustrates chaining.

---

**Example 4.7**

```
For J = 1,64
  C(J) = A(J) + B(J)
  D(J) = C(J) * E(J)
Endfor
```

Here, the output of add functional unit is an input operand to the multiplication functional unit. With chaining, we do not have to wait for the entire array C to be computed before beginning the multiplication. As soon as C(1) is computed, it can be used by the multiply functional unit concurrently with the computation of C(2). That is, the two functional units form the stages of a pipeline as shown in Figure 4.12(a).

Assuming that all the operands are in vector registers, this computation if done without vectorization (i.e., no pipelining or chaining), requires (add: $64 * 8 = 512$ plus multiply: $64 * 9 = 576$) 1088 clock periods. It can be completed in (chain start-up time of 17 plus 63 more

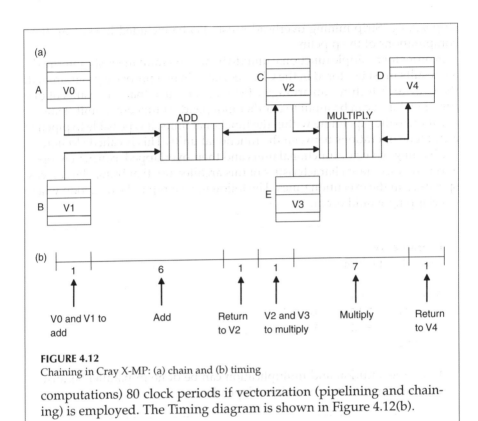

**FIGURE 4.12**
Chaining in Cray X-MP: (a) chain and (b) timing

computations) 80 clock periods if vectorization (pipelining and chaining) is employed. The Timing diagram is shown in Figure 4.12(b).

Note that the effect of chaining is to increase the depth of the pipeline and hence increase the start-up overhead.

If the operands are not already in vector registers, they need to be loaded first and the result stored in to the memory. The two load paths and the path that stores data to memory can be considered functional units in chaining. The start-up time for a load vector instruction is 17 cycles, and thereafter one value per cycle may be fetched, then any operation using this data may access one value per cycle after 18 cycles. Figure 4.13 shows an example of this. Here, port A is used to read in V0 and port B to read in V1. This occurs in parallel. As soon as each vector register has its first operand, the floating-point add may begin processing, and as soon as the first operand is placed in V2, port C may be used to store it back to memory.

In a chain, a functional unit can only appear once. Two fetches and one store are possible in each chain. This is because, Cray systems supply only one of each of the above types of functional units. This demands that if two floating-point adds are to be executed, they must occur sequentially. Because there are two ports for fetching vectors and one port for storing vectors, the user may view the system as having two load functional units and a store functional unit on Cray X-MP. On Cray-1, there is only one memory functional unit.

# Vector Processors

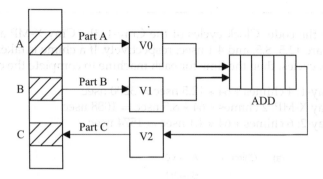

**FIGURE 4.13**
Memory and functional unit chaining on Cray X-MP

An operand can only serve as input to one arithmetic functional unit in a chain. An operand can, however, be input to both inputs of a functional unit requiring two operands. This is because a vector register is tied to a functional unit during a vector instruction. When a vector instruction is issued, the functional unit and registers used in the instruction are reserved for the duration of the vector instruction.

Cray Research has coined the term "chime" (chained vector time) to describe the timing of vector operations. A chime is not a specific amount of time, but rather a timing concept representing the number of clock periods required to complete one vector operation. This equates to length of a vector register plus a few clock periods for chaining. For Cray systems, a chime is equal to 64 clock periods plus a few more. A chime thus, is a measure that allows the user to estimate the speed-up available from pipelining, chaining, and overlapping instructions.

The number of chimes needed to complete a sequence of vector instructions is dependent on several factors. Since there are three memory functional units, two fetches and one store operation may appear in the same chime. A functional unit may be used only once within a chime. An operand may appear as input to only one functional unit in a chime. A store operation may chain onto any previous operation.

### Example 4.8

Figure 4.14 shows the number of chimes required to perform the following code on Cray X-MP, Cray-1, and the Cray-2 [Levesque and Williamson, 1989] systems:

```
For I = 1 to 64
   A(I) = 3.0 * A(I) + (2.0 + B(I)) * C(I)
Endfor
```

The Cray X-MP requires only two chimes, while the Cray-1 requires 4, and the Cray-2, which does not allow chaining, requires 6 chimes to execute the code. Clock cycles of the Cray-1, the Cray X-MP and the Cray-2 are 12.5, 8.5, and 4.1 nsec, respectively. If a chime is taken to be 64 clock cycles, then the time for each machine to complete the code is:

Cray-1: 4 chimes * 64 * 12.5 nsec = 3200 nsec
Cray X-MP: 2 chimes * 64 * 8.5 nsec = 1088 nsec
Cray-2: 6 chimes * 64 * 4.1 nsec = 1574 nsec

(a) Chime 1:   A → V0
               B → V1
               2.0 + V1 → V3
               3.0 * V0 → V4

    Chime 2:   C → V5
               V3 * V5 → V6
               V4 + V6 → V7
               V7 → A

(b) Chime 1:   A → V0
    Chime 2:   B → V1
               2.0 + V1 → V3
               3.0 * V0 → V4
    Chime 3:   C → V5
               V3 * V5 → V6
               V4 + V6 → V7
    Chime 4:   V7 → A

(c) Chime 1:   A → V0
    Chime 2:   B → V1
               3.0 * V0 → V4
    Chime 3:   C → V5
               2.0 + V1 → V3
    Chime 4:   V3 * V5 → V6
    Chime 5:   V4 + V6 → V7
    Chime 6:   V7 → A

**FIGURE 4.14**
Chime characteristics: (a) Cray X-MP, (b) Cray-1, and (c) Cray-2

# Vector Processors

Thus, for some instruction sequences, the Cray X-MP with the help of chaining can actually outperform Cray-2 which has a faster clock. Since Cray-2 does allow overlapping, the actual gain of Cray X-MP may not be as large for very large array dimensions.

During vector operations, up to 64 target addresses could be generated by one instruction. If a cache were to be used as intermediate memory, the overhead to search for 64 addresses would be prohibitive. Use of individually addressed registers eliminates this overhead. One disadvantage of not using a cache is that the programmer (or the compiler) must generate all the references to the individual registers. This adds to the complexity of code (or compiler) development.

### 4.3.3.1 Instruction fetch

Control registers are part of the special purpose register set and are used to control the flow of instructions. Four instruction buffers, each containing 32 Words (128 parcels, 16 bits each) are used to hold instructions fetched from memory. Each instruction buffer has its own instruction buffer address register (IBAR). The IBAR serves to indicate what instructions are currently in the buffer. The contents of IBAR are the high order 17 bits of the words in the buffer. The instruction buffer is always loaded on a 32-word boundary. The P register is the 24-bit address of the next parcel to be executed. The current instruction parcel (CIP) contains the instruction waiting to issue and the next instruction parcel (NIP) contains the next instruction parcel to issue after the parcel in the CIP. Also, there is a last instruction parcel (LIP) which is used to provide the second parcel to a 32-bit instruction without using an extra clock period.

The P register contains the address of the next instruction to be decoded. Each buffer is checked to see if the instruction is located in the buffers. If the address is found the instruction sequence continues. However, if the address is not found, the instruction must be fetched from memory after the parcels in the CIP and NIP have been issued. The least recently filled buffer is selected to be overwritten, so that the current instruction is among the first eight words to be read. The rest of the buffer is then filled in a circular fashion until the buffer is full. It will take three clock pulses to complete the filling of the buffer. Any branch to an out-of-buffer address causes a 16-clock pulse delay in processing.

Some buffers are shared between all of the processors in the system. One of these is real time clock. Other registers of this type include a cluster consisting of 48 registers. Each cluster contains 32 (1-bit) semaphore or synchronization registers and eight 64-bit ST or shared-T registers, and eight 24-bit SB or shared-B registers. A system with two processors will contain three clusters while a four-processor system will contain five clusters.

### 4.3.4 I/O System

The input and output of the X-MP is handled by the I/O subsystem (IOS). The IOS is made of 2 to 4 interconnected I/O processors. The IOS receives data from four 100 MB sec$^{-1}$ channels connected directly to the main memory of the X-MP. Also, four 6 MB sec$^{-1}$ channels are provided to furnish control between the CPU and the IOS. Each processor has its own local memory and shares a common buffer. The IOS supports a variety of front-end processors and peripherals such as disk drives and drives.

To support the IOS, each CPU has two types of I/O control registers: current address and channel limit registers. The current address registers point to the current word being transferred. The channel limit registers contain the address of the last word to be transferred.

### 4.3.5 Other Systems in the Series

Cray Research continued the enhancement of X-MP architecture to Y-MP and Y-prime series, while the enhancement of Cray-2 line was taken over by Cray Computer Corporation to develop Cray-3 and Cray-4 architectures. The Y-MP was of the same genre as the X-MP. The Cray-3 is a progression of Cray-2 architecture and was the first major system built with Gallium Arsenide (GaAs) technology. The Cray-3 was 10 times faster and less than half of the size of the Cray-2. The Y-MP was designed to appear as an X-MP to the user. It extended the X-MP 24-bit addressing scheme to 32 bits, thus allowing an address space of 32 million 64-bit words. It used a 6-nsec clock and eight processors, thus doubling the processing speed. The Cray-3 was a 16 processor machine offering a speed of 16 GFLOPS. The Y-prime series utilized silicon VLSI to improve on Y-MP while the Cray-4, a GaAs machine would use 64 processors to achieve a speed of 128 GFLOPS. Refer to Cray computer's website for details on the latest products.

## 4.4 Performance Evaluation

The major characteristics that affect supercomputer performance are:

- Clock speed
- Instruction issue rate
- Size and number of registers
- Memory size
- Number of concurrent paths to memory
- Ability to fetch/store vectors efficiently
- Number of duplicate arithmetic functional units (vector pipelines)
- Whether function calls can be "chained" together
- Indirect addressing capabilities
- Handling of conditional blocks of code

The high performance of all vector architectures can be attributed to the architectural characteristics similar to the above list. That is:

1. Pipelined functional units
2. Multiple functional units operating in parallel
3. Chaining of functional units
4. Large number of programmable registers
5. Block load and store capabilities with buffer registers
6. Multiple CPUs operating in parallel in a coarse-grained parallel mode
7. Instruction buffers

Comparison of vector processor performance is usually done based on *peak* and *sustained* computation rates in terms of MFLOPS. The peak performance rate can only be obtained under the most ideal circumstances and hence it does not reflect the true operating environment. Thus, the comparison based on sustained rates is more valid. The sustained performance is difficult to determine since it depends on a number of factors:

1. Level of vectorization, that is, the fraction of the application code that is vectorizable
2. Average vector length
3. Possibility of vector chaining
4. Overlap of scalar, vector, and memory load/store operations possible
5. Memory contention resolution mechanisms adopted

Consider an application program P. Let $T_{scalar}$ be the time to execute P entirely in scalar (or serial) mode. Let $s$ be the ratio of the speed of the vector unit to that of the scalar unit. Then, the time to execute P entirely in vector mode is $T_{scalar}/s$. If $f$ is the fraction of P that can be executed on the vector unit, the actual execution time for P in a mixed scalar/vector mode is

$$T_{actual} = (1-f)T_{scalar} + f \cdot T_{scalar}/s \qquad (4.4)$$

Hence, the speedup is

$$S = T_{scalar}/T_{actual}$$

$$= \frac{1}{(1-f) + f/s} \qquad (4.5)$$

This is known as Amdhal's law (Lubeck et al., 1985).

If $f$ is 1, the speedup equals $s$ and for $f = 0$, the speedup is 1. A significant increase in speedup is attained only when $f$ is closer to 1. That is, unless

**TABLE 4.2**

Peak throughputs of representative supercomputers

| Processor | MFlops | Cycle time |
|---|---|---|
| Cray X-MP/4 | 1100 | 8.5 |
| Cray-2 | 1800 | 4.1 |
| Cyber-205 | 800 | 20.0 |
| Hitachi S-810 | 630 | 4.5 |
| Fujitsu VP-200 | 533 | 15.0 |
| NEC SX-2 | 1300 | 6.0 |

a program is very heavily vectorized, the slower scalar mode of operation dominates the performance.

It is important to note that the Cray series of machines attacked Amdhal's law, by improving the scalar performance alone. In addition to providing an outstanding vector performance, the Cray-I was also the fastest scalar processor of its time.

Sustained performance of high-performance processors is measured by benchmarks that reflect accurately the types of workloads expected on these processors.

The peak throughput of representative supercomputers is shown in Table 4.2. In comparison, an Intel 80387 based personal computer with 20 MHz clock has a throughput of 0.44 MFLOPS. Section 4.6 provides throughput details of latest supercomputers.

Peak and sustained performance characteristics can be estimated based on the execution time of vector loops given in equation (4.1). This equation can be generalized to include other overheads as shown below:

$$T_N = T_{memory} + (T_{start\text{-}up} + T_{loop})(N/M + 1) + (N - 1)T_{cycle} \quad (4.6)$$

where $T_N$ is the execution time of a vector loop with $N$ elements, $T_{memory}$ the time to initialize starting addresses for each vector (a one-time overhead for the loop), $T_{start\text{-}up}$ the start-up time (including chaining), $T_{loop}$ the loop management overhead, $(N/M+1)$ the strip mining overhead, and $T_{cycle}$ is the pipeline cycle time (i.e., time to produce one result after the pipe line is full).

The peak performance measure ignores the overhead and hence is dependent entirely on $T_{cycle}$. If the loop contains $F$ floating-point operations, the peak rate in MFLOPS is given by:

$$(F/T_{cycle}) * \text{clock rate}$$

If the MFLOPS rating of a vector of length is denoted as $R_N$, the peak rate can be considered the rate for infinitely long vectors (i.e., $R$), which is given by:

$$R_\infty = \lim_{N \to \infty} \frac{F * \text{clock rate}}{\text{Clock cycles/iteration}} \quad (4.7)$$

The numerator is independent of $N$ and the denominator corresponds to $\lim(T_N/N)$ as $N \to \infty$.

Two other measures of performance commonly used are (Hockney and Jesshope, 1988) $N_{1/2}$ and $N_v$. $N_{1/2}$ is the vector length needed to achieve $R_\infty/2$. This is representative of the effect of start-up overhead, since in practice, vectors are not of infinite length. $N_v$ is the length of vector that makes the vector mode of computation faster than the corresponding computation in scalar mode. This measure takes into account the start-up overhead and also the speed of scalar operations relative to vector operations.

## 4.5 Programming Vector Processors

Vector processors tend to be difficult to program, just as any processor whose processing power comes from its unique hardware configuration. The hardware structures that make vector processors powerful also make the assembler coding difficult. If the machine is difficult to program at the assembler level, writing a compiler to take advantage of the hardware will not be any easier.

Two approaches have been used in providing programming facilities for parallel processors in general, and vector processors in particular: development of languages (or extensions to existing languages) that can express the parallelism inherent in the algorithm (Andrews and Olsson, 1993; Decyk, 1993) and development of compilers that recognize the portions of sequential code and "vectorize" them (Banerjee et al., 1993; Kuck, 1978; Polychronopoulos, 1988). Both the approaches have met partial success. In practice, these approaches complement each other rather than being exclusive of each other, since using only one of the two approaches may not result in the most efficient code.

In general, it is not possible to completely vectorize a sequential program. Those portions of the code that are not vectorizable will have to be executed in scalar hardware at a slower rate. Compiler writers in their quest to take advantage of the vector hardware as much as possible have tried to expand the realm of what is vectorizable. By developing new algorithms and approaches they have been able to take code segments that in the past have been executed in scalar mode and manipulate them so that they can be vectorized. In general, an algorithm that is considered efficient for scalar computation need not be efficient for vector environment. Modifications are then needed to take advantage of the vector hardware.

> **Example 4.9**
> Consider the following computation for a scalar processor (Ortega, 1988):
>
> ```
> y = 0
> for i = 1 to N
>     y = y + x(i) * a(i)
> endfor
> ```
>
> This can be altered for a vector processor as following:
>
> ```
> 0         →    r1
> a(1)      →    r2
> x(1)      →    r3
> r2 * r3   →    r4
> r1 = r1 + r4
> r1   →    main memory (y)
> y    →    r1  a(2)    →    r2
> .........
> (and so on)
> ```

Here, r1, r2, etc. are processor registers and → indicates a data transfer. The efficiency of this algorithm could be greatly improved if chaining is allowed by the vector processor and if we assume that most register-oriented vector processors allow load operations to occur during arithmetic operations, as shown below:

```
y        →    r1
a(1)     →    r2
x(1)     →    r3
y = y + r2 * r3 by chaining
a(2),x(2)    →    r1, r2
y = y + x(2) * a(2) by chaining
a(3),x(3)    →    r1, r2
......
(and so on)
```

This version of the algorithm offers a considerable performance increase over its predecessor. It requires two vector registers, one for $a$ and the other for $x$. It loads $a(i+1)$ while the multiplication of $x(i)$ and $a(i)$ is being performed, as well as loads $x(i+1)$ during the addition of $y$ to $x(i)a(i)$. It switches back and forth between the vector registers in order to provide the correct values for each operation.

# Vector Processors

Compilers do not always recognize the type of optimizations illustrated above, to take advantage of them. Nevertheless, there are several vectorization techniques adopted by vector processor environments to achieve high performance. Some of these techniques are described below.

*Scalar renaming*: It is typical for programmers to use a scalar variable repeatedly as shown by the following loop:

---

**Example 4.10**
```
for i=1,n
    x = A[i] + B[i]
         .
         .
    Y[i] = 2 * x
         .
         .
         .
    x = C[i]/D[i]
         .
         .
    P = x + 2
         .
         .
endfor
```

If the second instance of *x* is renamed as shown below, the two code segments become data independent, thus allowing a better vectorization.

```
for i = 1,n
    x = A[i] + B[i]
         .
         .
    Y[i] = 2 * x
         .
         .
         .
    xx = C[i]/D[i]
         .
         .
    P = xx + 2
         .
         .
endfor
```

*Scalar expansion*: In the following code segment, *x* is assigned a value and then used in a subsequent statement.

---

**Example 4.11**
```
for i = 1,n

    x = A[i] + B[i]
         .
         .
    Y[i] = 2 * x

endfor
```

If the scalar *x* is expanded into a vector as shown below, the two statements can be made independent, thus allowing a better vectorization.

```
for i = 1,n

    x[i] = A[i] + B[i]
         .
         .
    Y[i] = 2 * x[i]

endfor
```
---

*Loop unrolling*: For a loop of small vector length it is more efficient to eliminate the loop construct and expand out all iterations of the loop.

---

**Example 4.12**
The loop:
```
for I = 1 to 3

    x[I] = a[I] + b[I]

endfor
```
is unrolled into the following:

```
x[1] = a[1] + b[1]

x[2] = a[2] + b[2]

x[3] = a[3] + b[3]
```
---

This eliminates the looping overhead and allows the three computations be performed independently. (In this case, the computations at each iteration are not dependent on each other. If this is not the case, the computations must be partitioned into nondependent sets.)

*Loop fusion or Jamming*: Two or more loops that are executed the same number of times using the same indices can be combined into one loop.

**Example 4.13**
Consider the following code segment:

```
for i = 1, n
    X[i] = Y[i] * Z[i]
endfor
for i = 1, n
    M[i] = P[i] * X[i]
endfor
```

Note that each loop would be equivalent to a vector instruction. X is stored back into the memory by the first instruction and then retrieved by the second. If these loops are fused as shown below, the memory traffic can be reduced:

```
for i = 1, n
    X[i] = Y[i] * Z[i]
    M[i] = P[i] + X[i]
endfor
```

This assumes that there are enough vector registers available in the processor to retain X. If the processor allows chaining, the above loop can be reduced to:

```
for i = 1, n
    M[i] = P[i] + Y[i] * Z[i]
        endfor
```

*Loop distribution*: If the loop body contains dependent (i.e., statements that are data dependent) and nondependent code, a way to minimize the effect of the dependency is to break the loop into two, one containing the dependent code and the other nondependent code.

*Force maximum work into inner loop*: Since maximizing the vector length increases the speed of execution, the inner loop should always be made the longest. Further, dependency conflicts are avoided by shifting dependencies in an inner loop to an outer loop, if possible.

In addition to the above vectorization techniques, several techniques to optimize the performance of the scalar portions of the code have been devised. Some of these are described below.

*Subprogram in-lining*: For small subprograms, the overhead of control transfer takes longer than the actual subprogram execution. Calling a subprogram might consume about 10–15 clock cycles when no arguments passed and one argument might nearly double that overhead. In such cases, it is better to move the subprogram code into the calling program.

*Eliminate ambiguity using the PARAMETER statement*: If the (FORTRAN) Parameter statement is used to define a true constant in every subprogram in which it is used rather than passing it as a call-list parameter or putting it in a common block, the compiler can generate better optimized and unconditionally vectorized code.

*Positioning frequently executed scalar conditional blocks first*: Code in scalar conditional blocks (if-then-else) is executed in sequential order. Placing the most frequently executed conditional block early in the sequential code will speed up execution time of the code.

Construction of vectorizing compilers and languages for parallel processing has been an active area of research. A complete discussion of these topics is beyond the scope of this book. References listed at the end of this chapter provide further details.

Vector processor manufacturers provide various tools to assist the process of vectorization and multitasking. For instance, the LOOPMARK from Cray research Inc. shows which loops were vectorized and which were not. The FLOWTRACE utility measures each routine in the program in terms of time of execution, number of times called, calling routine, routines called percentage of overall execution time spent in routine. This information targets areas of code for revision and performance enhancement. Cray also provides a vectorizing FORTRAN compiler (CFT). Along with vectorization, multitasking also needs to be optimized. Cray provides a preprocessor for its CFT77 advanced FORTRAN compiler to aid in partitioning the code for optimal task dispersal.

## 4.6 Example Systems

This section provides a brief description of three supercomputer systems: Hitachi Super Technical Servers, NEC-SX, and Cray X1. The details in this section are extracted from respective manufacturer's literature.

### 4.6.1 Hitachi Super Technical Server

The Hitachi Super Technical Server series consists of SR8000, 10000 and 11000. Figure 4.15 shows the hardware configuration of the Hitachi SR8000 which was released in May 1998. It is a distributed-memory parallel

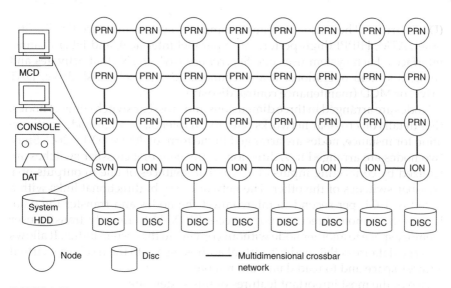

**FIGURE 4.15**
Hardware configuration of the SR8000 (Courtesy of Hitachi Ltd.)

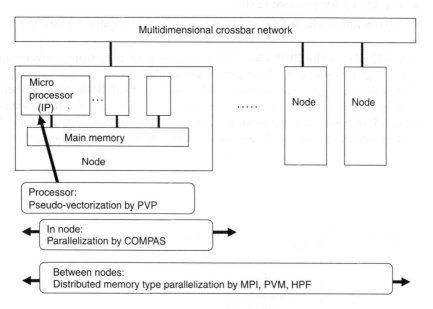

**FIGURE 4.16**
System architecture (Courtesy of Hitachi Ltd.)

computer consisting of an arbitrary number of nodes between 4 and 128. The nodes are interconnected by a multidimensional crossbar network. Each node consists of multiple microprocessors (IPs) that have shared memory (Figure 4.16). There are three types of nodes:

*Processing Nodes* (PRNs) are mainly used for performing calculations. *I/O Nodes* (ION) connect I/O devices such as Disk drives and Digital Audio Tape

(DAT) drives. They also offer small computer system interface (SCSI), Ethernet, ATM, HIPPI (high-performance parallel interface), and Fiber channel interfaces. Each system uses one *Supervisory Node* (SVN) that activates and controls system operations. It also connects to a console, disk drive, DAT drive, or MCD (maintenance control display).

Nodes are arranged in three dimensions and crossbar switch connect nodes. (For details on crossbar networks refer to Chapter 5.) In a 128-node configuration for instance, nodes are arranged in the form of 8(X) × 8(Y) × 2(Z). Since two nodes are arranged in Z direction, equivalent of 2 crossbar switches are realized by connecting inputs of Y crossbar swithces of one to outputs of Y crossbar switches of the other. The network uses bi-directional links with a speed of 1 GB per second to interconnect the nodes and transfer the data. It uses the remote-direct memory access (RDMA) to directly transfer user memory space to another node without copying to the system buffer. It allows to write data from the sender's virtual address space to the receiver's virtual address space and to read data from remote.

Two of the most important features of this system are:

- Pseudo vector processing (PVP)
- Co-operative Micro-Processors in single Address Space (COMPAS)

Figure 4.17 shows the PVP. The PVP is used to load data from memory to cache or registers without interrupting the calculations.

This is done using two features: Pre-fetch and Pre-load. Both of these operations are applied in software, but have hardware support. Pre-fetch loads 128-byte data from main memory to cache. Prefetch instructions are issued far before arithmetic instructions which use the data fetched by them. When

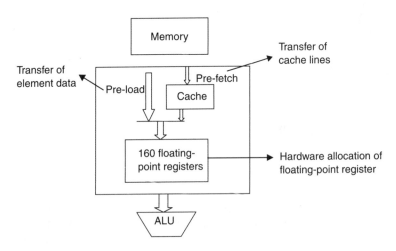

**FIGURE 4.17**
Pseudo vector processing (Courtesy of Hitachi Ltd.)

Vector Processors    155

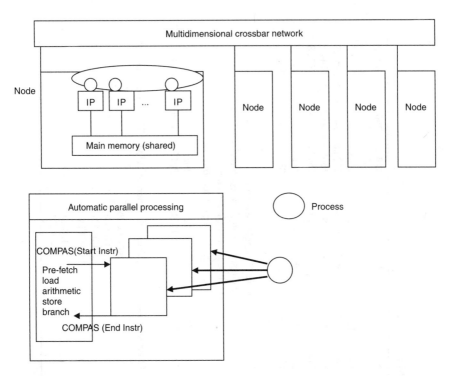

**FIGURE 4.18**
COMPAS (Courtesy of Hitachi Ltd.)

data are processed, they have been already prefetched in the cache. As such, fast data processing is achieved. When the main memory is accessed non-contiguously, pre-fetch is not very efficient because the prefetched cache line may contain unnecessary data.

To improve this situation the pre-load mechanism is implemented. Pre-load transfer data directly from main memory to registers bypassing the cache. When combined, pre-fetch and pre-load decreases the processor's waiting time between operations.

The COMPAS feature shown in Figure 4.18, divides a problem automatically between several processors. The system is based on creating threads for loops that signals (only) the instruction address of the loop. The number of threads generated by the operating system is equivalent to the number of processors. The compiler automatically performs parallelization in the node.

The Hitachi SR8000 E1 and F1 models are almost equal to the basic SR8000 model, except that the clock cycles for these models are 3.3 and 2.66 nsec, respectively. Moreover, the E1, F1, and G1 models can house more memory per node and the maximum configurations can be extended to 512 processors. The Hitachi has a bandwidth of 1.2 GB sec$^{-1}$ for the network in the E1 model while it is 1 GB sec$^{-1}$ for the basic SR8000 and the F1. By contrast, the G1 model has a bandwidth of 1.6 GB sec$^{-1}$. The characteristics of the different models are shown in Table 4.3.

**TABLE 4.3**
Specification of Hitachi SR8000 family (Courtesy of Hitachi Ltd.)

| System | | | 4 | 8 | 16 | 32 | 64 | 128 | 256 | 512 |
|---|---|---|---|---|---|---|---|---|---|---|
| | Number of nodes | SR8000 | | | | | | | | |
| | Peak | SR8000 | 32 | 64 | 128 | 256 | 512 | 1024 | — | — |
| | performance | SR8000 Model E1 | 38.4 | 76.8 | 153.6 | 307.2 | 614.4 | 1228.8 | 2457.6 | 4915.2 |
| | (GFLOPS) | SR8000 Model F1 | 48 | 96 | 192 | 384 | 768 | 1536 | 3072 | 6144 |
| | | SR8000 Model G1 | 57.6 | 115.2 | 230.4 | 460.8 | 921.6 | 1843.2 | 3686.4 | 7372.8 |
| | Inter-node | SR8000 | 1D | 2D | 2D | 2D | 3D | 3D | — | — |
| | crossbar | SR8000 Model E1/F1/G1 | 1D | 2D | 2D | 2D | 3D | 3D | 3D | 3D |
| | network | | | | | | | | | |
| | Inter-node | SR8000 | \multicolumn{8}{l}{1 GB sec$^{-1}$ (single direction)} | | | | | | | | |
| | transfer speed | SR8000 Model E1 | \multicolumn{8}{l}{1.2 GB sec$^{-1}$ (single direction) X2} | | | | | | | | |
| | | SR8000 Model F1 | \multicolumn{8}{l}{1 GB sec$^{-1}$ (single direction) X2} | | | | | | | | |
| | | SR8000 Model G1 | \multicolumn{8}{l}{1.6 GB sec$^{-1}$ (single direction) X2} | | | | | | | | |
| | Maximum total | SR8000 | 32 | 64 | 128 | 256 | 512 | 1024 | — | — |
| | memory capacity (GB) | SR8000 Model E1/F1/G1 | 64 | 128 | 256 | 512 | 1024 | 2048 | 4096 | 8192 |
| | External interface | | \multicolumn{8}{l}{Ultra SCSI, Ethernet/Fast ethernet, Gigabyte ethernet, ATM, HIPPI, Fiber channel} | | | | | | | | |
| Node | Peak | SR8000 | \multicolumn{8}{l}{8 GFLOPS} | | | | | | | | |
| | performance | SR8000 Model E1 | \multicolumn{8}{l}{9.6 GFLOPS} | | | | | | | | |
| | | SR8000 Model F1 | \multicolumn{8}{l}{12 GFLOPS} | | | | | | | | |
| | | SR8000 Model G1 | \multicolumn{8}{l}{14.4 GFLOPS} | | | | | | | | |
| | Memory | SR8000 | \multicolumn{8}{l}{2, 4, 8 GB} | | | | | | | | |
| | Capacity | SR8000 Model E1/F1/G1 | \multicolumn{8}{l}{2–16 GB} | | | | | | | | |

Hitachi SR8000 provides a set of parallel programming tools based on the Unix OS. HI-UX/MPP is the operating system for SR8000. It is a micro kernel based UNIX used for parallel distributed processing. It has 64-bit addressing with 32-bit code support, and scales from 4 nodes to the maximum of 512 nodes.

Hitachi SR8000 supports C, C++, Fortran 77, Fortran 90 which includes Fortran 95 specifications and Parallel Fortran (based on HPF version 2.0). The system also supports MPI-2 (Message Passing Interface version 2) and PVM (Parallel Virtual Machine) parallel development environments. Both of these can utilize the RDMA feature for fast communication between processes. Job scheduling is done in batch mode with NQS (Network Queuing System).

### 4.6.2 NEC SX Series

The NEC SX series has progressed from SX-6 to SX-8 models. SX-6 is a scalable parallel-type supercomputer designed for scientific and engineering computations in applications such as weather forecasting, molecular science, nuclear power, atomic fusion, structural analysis, and hydraulic analysis. The main system features are:

- Parallel-type supercomputer offering the performance range of 8 GFLOPS to 8 TFLOPS. A single-node system with 8 CPUs and 64-GB memory offers 64-GFLOPS and a 128-node system with 1024 CPUs with 8-TB memory offers 8-TFLOPS capability.
- Utilizes a UNIX-based Operating System (SUPER-UX) that supports POSIX interface as API and as a command interface.
- Provides software development environments to support both vector processing and parallel processing, by both shared and distributed memory configurations.

The Performance of the model group SX-6/M is shown in Table 4.4.

Figure 4.19 shows the system configuration of the SX-6 series. Two models of single-node system are possible: Model A with up to 8 CPUs and model B with up to 4 CPUs. Model A offers 16–64 GFLOPS vector performance. A multi-node system is configured with 2–128 single-node systems connected via an inter-node crossbar switch. In this type of system, users can operate a maximum of 128 nodes under the single system image. This distributed shared memory architecture built up of 1024 CPUs achieves a vector performance of 8 TFLOPS and provides a distributed and shared memory of 8 TB. The SX-6 series pursues the total balance of its processing performance, memory throughput as well as I/O performance.

The CPU performs instruction processing and especially executes programs consisting mainly of vector operations at high speed. For this purpose, the CPU of the SX-6 series is equipped with a vector unit that consists of a vector register and pipelines, each of which is used for logical operations,

**TABLE 4.4**

Performance of SX-6/M (Copyright © NEC Corporation, 2001)

| Model group | | SX-6/M | | |
|---|---|---|---|---|
| Model name | 64M8 | 32M4 | 16M2 | 8M2 |
| Central processing unit | | | | |
| Number of CPUs | 64 | 32 | 16 | 8 |
| Max. vector performance (GFLOPS) | 512 | 256 | 128 | 64 |
| Vector registers | 144 KB × 64 | 144 KB × 32 | 144 KB × 16 | 144 KB × 8 |
| Scalar registers | 64 bits × 128 × 64 | 64 bits × 128 × 32 | 64 bits × 128 × 16 | 64 bits × 128 × 8 |
| Main memory unit | | | | |
| Memory architecture | | Shared and distributed memory | | |
| Capacity (GB) | 512 | 256 | 128 | 128 |
| Max. data transfer rate | 2 TB sec$^{-1}$ | 1 TB sec$^{-1}$ | 512 GB sec$^{-1}$ | 256 Gbytes sec$^{-1}$ |
| I/O processor | | | | |
| Number of IOPs | 32 | 16 | 8 | 8 |
| Max. number of channels | 1016 | 508 | 254 | 254 |
| Max. data transfer rate (GB sec$^{-1}$) | 64 | 32 | 16 | 16 |
| Internode crossbar switch (IXS) | | | | |
| Max. data transfer rate (GB sec$^{-1}$) | 64 | 32 | 16 | 16 |

multiplication, add/shift operations, division, masked operations, and memory load/store. A single processor has eight sets of multiple parallel vector pipelines to achieve peak vector performance. In addition, the CPU is equipped with a large number of registers for scalar arithmetic operations and index calculations so that scalar arithmetic operations can be performed effectively. There are 128 64-bit scalar registers for storing scalar data. The scalar registers are used as base and index registers for address calculations, and as operand registers for fixed-point, floating-point, and logical operations.

The MMU is used to store user programs to be executed on the arithmetic processors and related data. The I/O processor (IOP) controls data transfers between the MMU and each of its peripheral devices, utilizing high-speed HIPPI and Gigabit Ethernet interfaces. Thus the SX-6 Series can be connected to a wide range of peripheral devices such as SCSI Disk Unit, Disk Array Unit (RAID), and SCSI Autoloader Tape Unit, etc.

The Remote Control Unit has an address translation device for internode data transfer and controls data transfer to the remote node, global communication register access, and inter-unit communication corresponding to requests from the CPU.

# Vector Processors

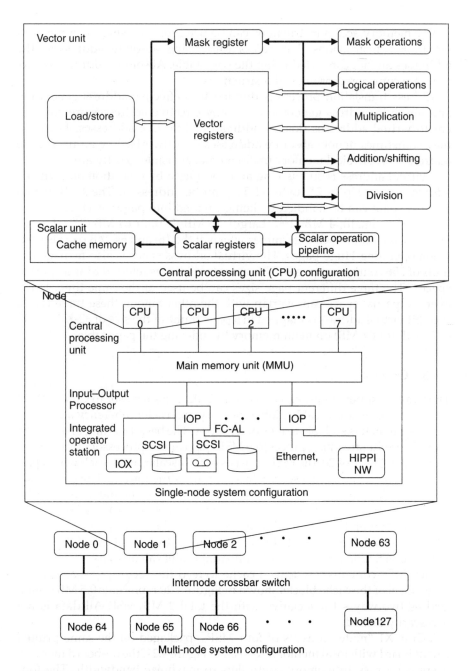

**FIGURE 4.19**
System configuration of NEC-SX6 (Copyright © NEC Corporation, 2001.)

Any memory access instructions can be used for accessing main memory (MM). For instructions that access MM using absolute addresses, the addresses are not converted using the page table. Absolute Address instructions are considered privileged instructions.

Address translation process is divided into effective address generation and absolute address generation. Usually effective addresses are equivalent to virtual addresses (logical addresses). Effective addresses, however, may sometimes denote absolute addresses directly. Absolute addresses are addresses on physical memory (main memory) visible to software.

A virtual address is 40 bits long and specifies a byte position on a virtual address space. Up to $2^{40}$ bytes (1 TB) can be addressed. The 1-TB virtual address space can be handled by being divided into pages, each consisting 32 KB ($2^{15}$) (Small), 4 MB ($2^{22}$) (Large), 16 MB ($2^{24}$), or 64 MB ($2^{26}$) (Huge). System Generation (SG) determines which of the 16- and 64-MB page is used when operating Huge pages. The virtual address space is also managed in units of 256 contiguous pages. This managed space is referred to as a "partial space." The three different page sizes can be used at the same time. However, no partial space can overlap with the other among these three types. A 40-bit virtual address is mapped in a page frame of the size of 32-KB, 4-MB, or 16-MB (or 64-MB) on main memory by indexing the page table.

### 4.6.3 Cray X1

The Cray X1 supercomputer, launched in 2002, is a successor to Cray MPP. It offers a directly addressable global memory and cache coherency across the entire system. It is scalable to tens of teraflops. The basic building block of the X1 is the multi-streaming processor (MSP) just called "processor" by Cray. As depicted in Figure 4.20, the MSP consists of four scalar units, each with a pair of vector pipes. Each scalar section has its own instruction and data caches and can dispatch up to two instructions per clock and execute them out of order in one of its execution units. The four scalar units operate at 400 MHz and can deliver a total of 3.2 GFLOPS. Combined, the eight vector pipes running at 800 MHz can deliver 12.8 GFLOPS for 64-bit floating-point calculations and 25.6 GFLOPS when executing 32-bit floating-point calculations.

Each MSP contains interleaved cache to feed the vector pipes and scalar units. Each of the cache blocks shown in Figure 4.20 represents 0.5 MB organized as two-way set associative with the total 2 MB/MSP. All data is not necessarily cached.

A Cray X1 "node" consists of four MSPs, packaged on the same printed circuit board with local memory. As seen in Figure 4.21, the on-board memory is spread across 16 memory controllers to maximize bandwidth. The four cache chips on each MSP interleave across the 16 sets of memory chips in a way that reduces the complexity of the on-node interconnect, allowing it to retain low latency.

The two I/O channel pairs per node shown in Figure 4.21 provide connections to disk and other I/O by way of a PCI-X I/O subsystem. Most of

# Vector Processors

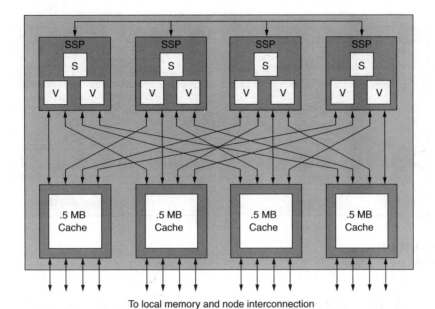

**FIGURE 4.20**
Basic multi-streaming processor (Courtesy of Cray Inc.)

the I/O connections are fiber channel based. Gigabit Ethernet, legacy HIPPI, and other protocols can be supported through the attached Cray Network Server (CNS).

The Cray X1 has either air-cooled or liquid-cooled cabinets. The air-cooled cabinets contain up to four nodes, or 16 MSPs. Radiator-like heat exchangers transfer heat from the closed-loop Fluorinet system to air being blown through the cabinet. Since a chilled water loop is able to remove heat more efficiently than air, the liquid-cooled cabinets allow up to 16 nodes or 64 MSPs (819.2 GFLOPS). In this case, a heat exchanger transfers heat from the Fluorinet directly to the customer-supplied chilled water. The large chiller units that cool the air in the computing center are typically water cooled.

The development environment for Cray X1 systems is the software that is provided with or that runs on top of the UNICOS/mp operating system. The UNICOS/mp operating system is based on the IRIX 6.5 operating system. A single UNICOS/mp kernel image runs on the entire system. The software comprises:

- Commands, system calls, system libraries, special files, protocols, and file formats provided with the UNICOS/mp operating system.
- Cray programming environment:
    a. Cray FORTRAN Compiler and the Cray C and C++ Compilers.

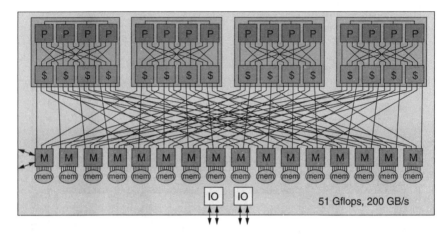

**FIGURE 4.21**
Node, containing four MSPs (Courtesy of Cray Inc.)

b. Cray Tools, which includes the Cray loader and CrayPat, the performance analysis tool for Cray X1 system.
c. Libraries for application programmers.
d. The Modules utility.
e. X Window System X11 libraries.

A Cray X1 system includes a CNS for network routing. In addition to bridging from Fiber Channel connections to network protocols, the CNS provides IP packet management, minimizing the impact on the Cray X1 processors. Initially, the CNS supported Gigabit Ethernet and HIPPI network types. Multiple CNSs can be configured on a Cray X1 system to scale the number of network connections required.

## 4.7 Summary

Architectural issues concerned with vector processing schemes are covered in this chapter. These architectures mix pipelining and parallel processing to various degrees. Basic models for the two types of vector processor architectures are provided along with the architectural highlights of one commercial vector processor system. Brief description of three contemporary systems is also provided.

Vector processors belong to the superpipelined architectures category and are suitable for applications that offer a high level of vectorization. These are the applications that call for very long vectors. Superscalar architectures on

*Vector Processors*

the other hand, issue multiple instructions per cycle. These are suitable for applications with a high degree of instruction level parallelism.

In general, specialized programming schemes and compilers are needed to efficiently utilize the computational power offered by these architectures. A brief discussion of the programming concepts and vectorization of serial programs is provided in this chapter.

With the advances in hardware technology, it is now possible to build very highspeed pipelined scalar processors. These are usually much cheaper to build and offer performance levels close to those of vector processor architectures, thereby making vector processor architectures obsolete. It is interesting to note that the performance of some modern workstations, notably those with good floating point hardware, is approaching the performance of supercomputers, at least in terms of MFLOPS. The major difference between the two, however, is in terms of the memory bandwidth. The workstations with their dynamic RAM main memories cannot compete in memory bandwidth, with high speed, highly interleaved and pipelined memory systems of supercomputers. Thus, the place where the supercomputers really shine is in operating on extremely large datasets, and not just in raw FLOPS.

## Problems

4.1. The following are common operations on matrices: column sum, row sum, transpose, inverse, addition, and multiplication. Examine the algorithm for each operation and develop vectorized procedures for each assuming a vector processor capable of performing vector add, subtract, reciprocal, and multiply operations.

4.2. The following code segment need to be executed on a vector processor with a five-stage floating-point adder and a six-stage floating-point multiplier:

$$A = B + C$$
$$D = s * A$$
$$E = D + B$$

where **A**, **B**, **C**, **D**, **E** are 32-element vectors and $s$ is a scalar. Assume that each stage of the adder and multiplier consumes one cycle, memory load, and store each consume one cycle and there are required numbers of memory paths. Derive the computation time if (1) chaining is not allowed (2) chaining is allowed. Show the timing diagram for each case.

4.3. Solve the above problem, assuming that only one path to and from memory is available.

4.4. Assume that the vector processor has vector registers that can hold 64 operands and data can be transferred between these registers in blocks of 64 operands every cycle. Solve Problem 4.2 with these assumptions.

4.5. Modify the following program segments for efficient execution on Cray architecture for $N = 16, 64,$ and 128:

(a)
```
for I = 1 to N
    for J = 1 to 20
        B(I,J) = K * A(I,J)
    endfor
endfor
```
(b)
```
for I = 1 to N
    X(I) = K * X(I)
    Y(I) = Y(I) + X(I-1)
endfor
```

4.6. For each loop in the above problem, find $R_{100}$ and $R_\infty$ and $N_{1/2}$. Assume $T_{memory} = 10$ cycles and $T_{loop} = 5$ cycles.

4.7. Consider the problem of sorting very large arrays of numbers. Investigate how a vector processor can be used effectively in this task. Select a sorting algorithm and vectorize it as far as possible.

4.8. Assume that a vector processor has 16 banks of memory organized as in Figure 4.4, with two-cycle access time. Investigate the best way to represent a $32 \times 32$ matrix, if the computation requires accessing:

(1) One row at a time
(2) one column at a time
(3) either a row or a column at a time

4.9. Design a memory structure suitable for accessing the matrix in the above problem for a vector processor with vector registers capable of holding 16 elements of data.

4.10. Develop an instruction set for a vector processor that is required to perform common matrix operations. Assume that (1) memory-oriented and (2) register-oriented models.

4.11. Compare the architecture and performance characteristics of contemporary processor/numeric coprocessor pairs with those of the commercial vector processor architectures described in this chapter.

4.12. Most vector processors use only an instruction cache and no data cache. Justify this architecture.

4.13. Compare two contemporary supercomputer families with respect to their instruction sets, instruction execution speeds, and policies. What support is provided by the compilers for these machines, to make the applications run faster?

4.14. Discuss the special array operations needed on vector processors to enable parallel execution of conditional operations, that is, vectorizing loops containing branches.

4.15. Assume that a vector processor operates 20 times faster on vector code than scalar code. If only $x\%$ of the program is vectorizable, what is the value of $x$, for the machine to execute the program twice as fast as the scalar processor?

## References

Andrews, G. R. and Olsson, R. A. *The SR Programming Language: Concurrency in Practice.* Redwood City, CA: Benjamin/Cummings, 1993.

Banerjee, U., Eigenmann, R., Nocilau, A., and Padua, D. A. Automatic program parallelization. *Proceedings of IEEE*, 81, 1993, 211–243.

Cray Inc. *Cray X1*™ *System Overview Manual*, S–2346–22, 2004. http://www.cray.com/.

Cray Research, Inc. *Cray X-MP and Cray Y-MP Computer Systems* (Training workbook), 1988.

Decyk, V. K., How to write (nearly) portable Fortran programs for parallel computers, *Computers in Physics*, 7(4), pp. 418–424, July/August 1993.

Hack, J. J. Peak vs. sustained performance in highly concurrent vector machines. *IEEE Computer*, 19(2), 1986, 11–19.

Hitachi. *SR8000 Series Super Technical Server*, http://www.hitachi.co.jp/Prod/comp/hpc/eng/sr81e.html

Hitachi. *SR8000 User's Guide*, 2000.

Hitachi. *SR11000*, http://www.hitachi.co.jp/Prod/Comp/hpc/SR_e/11ktop_e.html

Hockney, R. W. and Jesshope, C. R. *Parallel Computers*, Vol. 2. Philadelphia, PA: Adam Hilger, 1988.

Kuck, D. J. *The Structure of Computers and Computations.* New York: John Wiley & Sons, 1978.

Levesque, J. M. and Williamson, J. L. *A Guidebook to Fortran on Supercomputers.* San Diego, CA: Academic Press, 1989.

Lubeck, O., Moore, J., and Mendez, P. A benchmark comparison of three supercomputers: Fujitsu VP-200, Hitachi S810/20 and Cray X-MP/2. *IEEE Computer*, 18, 1985, 10–24.

NEC Solutions, *SX Series Supercomputers.* http://www.sw.nec.co.jp/hpc/sx-e/sx6/index.html and http://www.hpce.nec.com/

Ortega, J. M. *Introduction to Parallel and Vector Solution of Linear Systems.* New York: Plenum Press, 1988.

Polychronopoulos, C. D. *Parallel Programming and Compilers.* Boston, MA: Kluwer Academic Publishers, 1988.

Vaughan-Nichols, S. J. New trends revive supercomputing industry. *IEEE Computer*, 37(2), February 2004, 10–13.

# 5

# Array Processors

This chapter covers parallel array processors, which are the SIMD (single instruction stream, single data stream) class of architectures introduced in Chapter 2. Recall that the major characteristics of these architectures are:

1. A single control processor (CP) issues instructions to multiple arithmetic/logic processing elements (PEs).
2. PEs execute instructions in a "lock-step" mode. That is, PEs are synchronized at the instruction level and all the PEs execute the same instruction at any time and hence the name *synchronous array processors*.
3. Each PE operates on its own data stream. Since multiple PEs are active simultaneously, these architectures are known as *data-parallel* architectures.
4. These are *hardware intensive architectures* in the sense that the degree of parallelism possible depends on the number of PEs available.
5. The PEs are usually interconnected by a data exchange network.

The next section revisits the SIMD organization illustrated in Figure 2.5 as a means of developing a generic model for this class of architectures. The performance of an SIMD architecture is very much influenced by the data structures used and the memory organization employed. These topics are covered in Section 5.2. Section 5.3 introduces interconnection networks and provides details of commonly used topologies in SIMD architectures. Section 5.4 provides performance evaluation concepts. Section 5.5 discusses programming considerations. Section 5.6 describes two SIMD architectures: the ILLIAC-IV an experimental machine for its historical interest and Thinking Machine Corporation's Connection Machine (CM), a commercial system.

## 5.1 SIMD Organization

The following example compares the operation of an SIMD with that of an SISD (single instruction stream, single data stream) system.

**Example 5.1**
Consider the addition of two $N$-element vectors **A** and **B** element-by-element to create the sum vector **C**. That is,

$$C[i] = A[i] + B[i] \quad 1 \leq i \leq N \tag{5.1}$$

As noted in Chapter 4, this computation requires $N$ add times plus the loop control overhead on an SISD. Also, the SISD processor has to fetch the instructions corresponding to this program from the memory each time through the loop.

Figure 5.1 shows the SIMD implementation of this computation using $N$ PEs. This is identical to the SIMD model of Figure 2.5 and consists of multiple PEs, one CP and the memory system. The processor interconnection network of Figure 2.5 is not needed for this computation. The elements of arrays $A$ and $B$ are distributed over $N$ memory blocks and hence each PE has access to one pair of operands to be added. Thus, the program for the SIMD consists of one instruction:

$$C = A + B$$

This instruction is equivalent to

$$C[i] = A[i] + B[i] \quad 1 \leq i \leq N$$

where $i$ represents the PE that is performing the addition of the $i$th elements and the expression in parentheses implies that all $N$ PEs are active simultaneously.

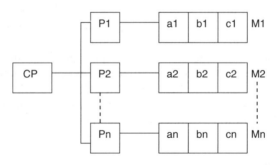

**FIGURE 5.1**
SIMD processing model

The total execution time for the above computation is the time to fetch one instruction plus time for one addition. No other overhead is needed. But the data needs to be structured in $N$ memory blocks to provide for the simultaneous access of the $N$ data elements.

*Array Processors*

Thus, SIMDs offer an $N$-fold throughput enhancement over the SISD provided the application exhibits a data-parallelism of degree $N$. The following sections examine in detail the functional requirements of each component of an SIMD.

### 5.1.1  Memory

In the SIMD model of Figure 5.1 and Figure 2.5, each PE is connected to its own memory block. The conglomeration of all the memory blocks forms the system memory. Programs typically occupy parts of several memory blocks. For instance, if the memory addresses are interleaved between the blocks, instructions in a program would spread among all the blocks of the memory. Data are distributed among the memory blocks such that each PE can access the data it operates on. The CP accesses the memory for instruction retrieval and PEs access the memory for data retrieval and storage. This simultaneous access requirement might result in memory access bottleneck.

One way of minimizing the memory access bottleneck is to include a memory block at the CP in which the instructions reside. This organization partitions the system memory into data and program memories. Since the complete memory system is now not available for program storage, the programs sizes are restricted by the program memory size.

In both of the above organizations, each PE has its own (i.e., local) memory. This requires that the data be loaded into the memory such that the data elements required for computation by a PE are available in its local memory. Another possible variation is where data memories are connected to PEs with an ($n \times n$) interconnection network (switch). This switch allows any data memory to be accessed by any PE. That is, the switch is a *data alignment* network. This organization allows a flexible data structuring at the cost of the switch complexity.

As we will see later in this chapter, data structuring is an important consideration in the application of SIMD architectures. In the absence of a data alignment network, additional operations are needed to rearrange data in the memory appropriately (i.e., to facilitate simultaneous access by PEs) as the computation progresses. For example, assume that data is originally organized to provide simultaneous access to the row elements of a matrix. If the subsequent computation requires simultaneous access to column elements, data needs to be reloaded in the memory appropriately before the new computation begins. Several data structuring techniques have been devised over the years to allow various modes of accessing data in SIMDs. Section 5.3 describes some of these techniques.

### 5.1.2  Control Processor

The CP fetches instructions and decodes them. It transfers arithmetic or logic instructions to PEs for execution. That is, the CP generates control signals corresponding to these instructions that are utilized by the PEs. Control instructions are executed by the CP itself. The CP performs all the address computations and might also retrieve some data elements from the memory and *broadcast* them to all the PEs as required. This method of data distribution can be employed when either a PE interconnection network is not available or broadcasting is more efficient than using the PE interconnection network. The CP is thus equivalent to an SISD processor except that the arithmetic/logic functions are not performed by it.

### 5.1.3 Arithmetic/Logic Processors

The PEs perform the arithmetic and logical operations on the data. Thus each PE corresponds to data paths and arithmetic/logic units (ALUs) of an SISD processor, capable of responding to control signals from the control unit.

### 5.1.4 Interconnection Network

Figure 5.2 shows three top level models for SIMD architectures (Skillicorn, 1985). The Instruction processor (IP) in these models is the same as the CP in the SIMD model of Chapter 2, and the Data Processor (DP) is same as the PE. All models have a 1-to-$n$ switch connecting the single IP with $n$ DPs. In Figure 5.2(a) there is a single memory hierarchy (i.e., common data and instruction memory (IM)). There is a 1-to-$n$ switch between the IP and the memory hierarchy, and the DP to memory switch is $n$-to-$n$. The DPs are connected by an $n$-by-$n$ switch. This corresponds to the SIMD model of Figure 2.5, which also depicts the structure of ILLIAC-IV. The DPs in ILLIAC-IV are actually connected by a torus, or mesh network in which there is no $n$-by-$n$ physical interconnection. But it is possible to exchange data between all the $n$ DPs by repeated use of the interconnection network, and hence, it is an $n$-by-$n$ switch. In Figure 5.2(b) and 5.2(c), separate IM and data memory (DM) hierarchies are shown. The IP to IM interconnection is thus 1-to-1.

In Figure 5.2(b), the DP interconnections are $n$-by-$n$ and the DP to DM interconnections are $n$-to-$n$. This model depicts the structure of Thinking Machine Corporation's CM. The CM uses a hypercube network between the DPs. It is not physically an $n$-by-$n$ interconnection scheme but allows data exchange between all $n$ DPs by repeated use of the network and hence logically is an $n$-by-$n$ switch.

In Figure 5.2(c), there is no direct connection between DPs and the interconnection between DM and DPs is $n$-by-$n$. This model depicts the structure of Burroughs Scientific Processor (BSP). The DP–DM switch in BSP is known as a data alignment network because it allows each DP to be connected to its data stream by dynamically changing the switch settings. Note that if the DP–DM switch is $n$-to-$n$, as in Figure 5.2(a) and 5.2(b), data has to be moved into appropriate memory elements to allow access from the corresponding DP.

# Array Processors 171

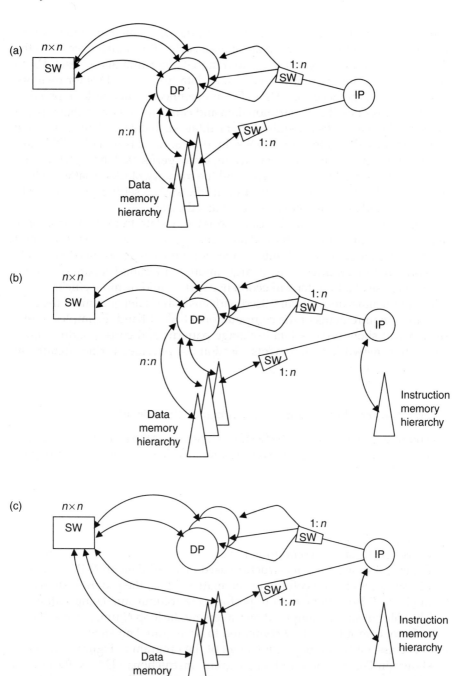

**FIGURE 5.2**
SIMD models (Skillicorn, 1985): (a) Type 1, (b) Type 2, and (c) Type 3 (Reproduced from IEEE. With permission.)

In Type 1 and Type 2 SIMD architectures, the DP to memory interconnection is through an $n$-to-$n$ switch and hence each DP has direct access to the data in its local memory. If the computational algorithm allows each DP to continue processing without needing any data from other DPs (or memory blocks), there is no need for a DP to DP interconnection network. In practice, it is not always possible to partition data and computations such that no data exchange between DPs is required. To facilitate such data exchange, an $n \times n$ switch was used to interconnect DPs. In Type 3, there is no DP to DP interconnection network since all data exchanges are performed through the $n \times n$ (alignment) switch between the DPs and the memory blocks. That is, instead of moving the data to the DP needing it, the switch settings are altered to connect the DP to the memory that contains the data.

As mentioned earlier, the CP can also retrieve data from memory blocks and distribute them to PEs. If the data exchange is not excessive this method is efficient enough and a PE interconnection network is not needed. But if the computation requires either a large amount or a complex pattern of data exchange, use of a PE interconnection network would be more efficient.

The interconnection network used with an SIMD is dependent on the data exchange requirements of the computational task at hand. Examples in this section will show typical data exchange patterns and corresponding interconnection network requirements. Section 5.4 provides further details on interconnection networks.

### 5.1.5 Registers, Instruction Set, Performance Considerations

The unique characteristic of the SIMD architecture is the lock-step instruction execution. This implies that the computational task must exhibit a high degree of parallelism (at the instruction level) and data must be structured to allow simultaneous access by the PEs.

---

**Example 5.2**
Consider the task of computing the column sum of an $(N \times N)$ matrix. Figure 5.3 shows the data structure used in an SIMD with $N$ PEs. Each PE has access to one column of the matrix. Thus, the program shown in Figure 5.3(b) can be used to perform the column sum computation. The column sum is computed by traversing the loop $N$ times. Thus, the order of computation is $N$ compared to $N^2$ required on an SISD.

The assembly language equivalent of the program in Figure 5.3(b) is shown in Figure 5.3(c). In this program, instructions LDA, ADD, and STA are executed simultaneously by all the PEs, while the instructions LDX, DEX, BNZ, and HLT are executed by the CP. Thus the instruction set of an SIMD is similar to that of an SISD except that the arithmetic/logic instructions are performed by multiple PEs simultaneously.

# Array Processors

(a)

| | M1 | M2 | M3 | ---- | Mn |
|---|---|---|---|---|---|
| | a11 | a12 | a13 | | a1n |
| | . | . | . | | . |
| | . | . | . | | . |
| | . | . | . | | . |
| | an1 | | | | ann |
| | sum1 | sum2 | | | sum n |
| | P1 | P2 | P3 | ---- | PN |

(b) Sum [i] = 0 ( $1 \leq i \leq N$ )
For j = 1 to N
  Sum [i] = Sum [ i ] + A [ i ] [ j ] ($1 \leq i \leq N$)
End for

(c)

```
              LDA       ZERO            load the accumulator with a 0.
              LDX       N               load index reg. with N.
      LOOP    ADD       A-1, 1
              DEX                       Decrement index
              BNZ       LOOP            Branch to loop if index  0
              STA       SUM             Store Sum
              HLT                       Halt
      ZERO    CONSTANT  0
      N       CONSTANT  5               For a 5×5 Matrix
      A       CONSTANT  4,2,3,5,6       Column elements
              END
```

**FIGURE 5.3**
Matrix column addition: (a) data structure, (b) high-level program, and (c) assembly-level program

A global index register that is manipulated by the CP to access subsequent elements of data in individual memory blocks, has been assumed in the above example. In practice, each PE will have its own set of (local) index registers in addition to global registers. This will allow a more flexible data storage and access. Correspondingly, the instruction set will contain two types of index manipulation instructions, one set for the global registers and the other for local registers.

### Example 5.3
Now consider the problem of normalizing each column of an ($N \times N$) matrix with respect to the first element of the column. This can be done in $N$ time units using an SIMD with $N$ PEs and the data structure is same as that in Figure 5.3. Since this computation requires a division, if the first element of any column is zero, that column should not participate in the

normalization process. Since the SIMD works in a lock-step mode, the PEs corresponding to the columns with a zero first element will have to be deactivated during this computation.

To facilitate this type of operation, each PE contains an activity bit as part of its status word. It is set or reset based on the condition within that processor's data stream and controls the activity of the PE during the subsequent cycles.

In addition to the usual instructions that set the condition code bits in the status register, the instruction set of an SIMD contains instructions that allow logical manipulation of the activity bits of all the PEs. These instructions enable manipulation of activity bits so as to accommodate the activation of appropriate PEs based on complex data conditions.

Conditional branch instructions in the program also utilize the activity bits of PEs as shown by the following example:

**Example 5.4**
Consider the program segment depicted by the flowchart of Figure 5.4. Here, A, B, C, and D represent blocks of instructions. If this program

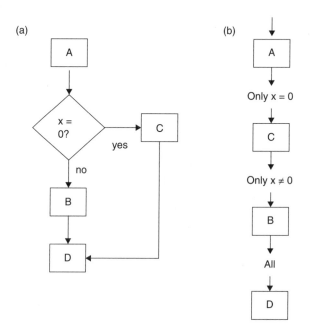

**FIGURE 5.4**
Conditional branch: (a) SISD version and (b) SIMD version

# Array Processors

> were for an SISD, the processor after executing A, would have executed either B or C depending on the value of X and then execute D as shown in Figure 5.4(a). In an SIMD, some data streams satisfy $X = 0$ and the others satisfy $X \neq 0$. That means, some PEs execute B and the others execute C and all PEs eventually execute D. To accommodate this computation and to retain the instruction lock-step mode of operation, the branch operation is converted into a sequential operation as shown in (b). Note that all the PEs are not active during the execution of blocks C and B and hence the SIMD hardware is not being utilized efficiently. As such, for an SIMD to be efficient, conditional branches should be avoided as far as possible.

Note that the PE interconnection network was not needed in the above applications. The examples in the next section illustrate the need for such a network and provide details of data storage concepts.

## 5.2 Data Storage Techniques and Memory Organization

As mentioned earlier, the data storage pattern has a significant effect on the performance of SIMDs. The following example introduces the two common data storage formats for matrices.

> **Example 5.5**
> Consider the *straight storage* format of matrices used in Figure 5.3. It allows accessing all the elements of a row of the matrix simultaneously by the PEs. Thus, while it is suitable for column oriented operations, it is not possible to access column elements simultaneously. If such an access is needed (for row oriented operations), the matrix storage needs to be rearranged such that each row is stored in a memory block (as shown in Figure 5.5). This rearrangement is time consuming and hence impractical especially for large matrices.
>
> The *skewed storage* shown in Figure 5.6, solves this problem by allowing simultaneous row and column accesses. Since the matrix is skewed, additional operations are needed to align the data elements. For instance, when accessing rows as shown in Figure 5.6(a), note that the elements in the first row are properly aligned. But the second row elements need to be rotated left once, after they are loaded into the processor registers. In general, the $i$th row elements need to be rotated $(i-1)$ times to the left to align them with the PEs.

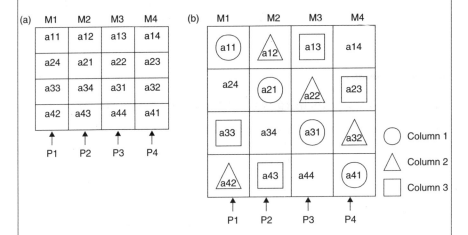

**FIGURE 5.5**
Row-wise matrix representation

**FIGURE 5.6**
Skewed matrix representation: (a) row access and (b) column access

Accessing of columns is shown in Figure 5.6(b). Note that this requires local index registers. To access the first column, the indexes are set as shown by arrows. To access subsequent columns, the indexes are decremented by 1 modulo $N$. After the elements are loaded into the PE registers, they are aligned by rotating them. The elements of $i$th column are rotated $(i - 1)$ times left.

The skewed storage thus necessitates a *ring* or *loop* interconnection network between the PEs. This network connects each PE to its right or left neighbor (unidirectional) or both (bidirectional). Correspondingly, the instruction set contains instructions to utilize the functions provided by the PE interconnection network. For instance, in the case of the loop interconnection

# Array Processors

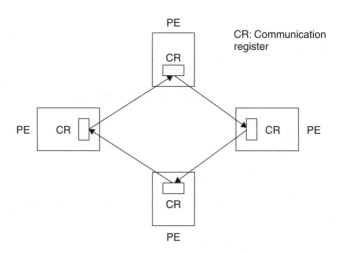

**FIGURE 5.7**
Schematic of ring interconnection network with four PEs

instructions of the type SHIFT LEFT and SHIFT RIGHT would be needed. These instructions when issued shift the data simultaneously between the PEs. The data transfer between the PEs over the interconnection network usually occurs through a set of registers within each PE. The results of computation are normally retained in these registers so that they can be shared with other PEs at register transfer speeds. Figure 5.7 shows a schematic of a unidirectional ring with four PEs.

In general an algorithm that is efficient for an SISD need not be efficient for an SIMD, as illustrated by the following example:

---

**Example 5.6**

Consider for instance, the popular matrix multiply algorithm for an SIMD depicted by the program in Figure 5.8(a). Here, to compute the element $C[i, j]$, the $i$th row elements of A are multiplied element-by-element by the $j$th column elements of B and the products are accumulated. The algorithm executes in $N^3$ time.

To implement this algorithm on an SIMD, the first attempt would be to store matrix A column-wise and B row-wise so that appropriate row and column can be accessed simultaneously. This allows the generation of the $N$ products simultaneously. But, since these products are in $N$ processors, they need to be accumulated sequentially by one processor. Thus, the computation is not going to be efficient.

A slight rearrangement of the algorithm yields a better SIMD computation. The SIMD program shown in Figure 5.8(b) assumes that all

```
(a)   for K = 1 to N
          for i = 1 to N
              C[ i , k ] = 0
              for j = 1 to N
                  C[ i , k ] = C[ i , k ] + A[ i , j ]*B[ j , k ]
              endfor
          endfor
      endfor

(b)   for i = 1 to N
          C [ i, k ] = 0   (1 ≤ k ≤ N)
          for j = 1 to N
              C[ i , k ] = C[ i , k ] + A[ i , j ] * B [ j , k]    (1 ≤ k ≤ N)
          endfor
      endfor
```

**FIGURE 5.8**
Matrix multiplication: (a) SISD version and (b) SIMD version

the three matrices are stored in column-wise straight storage format in $N$ memory banks. Here $k$ is the processor number. Elements of the $i$th row are computed one product at a time through the $j$ loop. It takes $N$ iterations of the $j$ loop to complete the computation of the $i$th row elements of C. This algorithm executes in $N^2$ time and does not require a sequential mode of operation on the part of PEs.

Note that within the $j$ loop, the element $A[i,j]$ is available at the $j$th processor and is required by all the other processors. This is done by *broadcasting* the content of this value from $j$th processor to others. Broadcast is a common operation in SIMDs and is implemented in one of the two ways mentioned earlier: either the PE interconnection network performs the broadcast or the CP reads the value from the $j$th PE and supplies it to all the PEs. For instance, if the ring interconnection network (IN) is used, $(N-1)$ shifts are needed to broadcast the value in any PE. Alternatively, the CP can read the value from the PE and broadcast it in one instruction cycle time.

The remainder of this section concentrates on the memory organization techniques that allow efficient data structuring on SIMDs. As shown above, if a matrix is to be accessed only by rows, we can store the matrix in row-wise straight storage format. If the columns of a matrix are to be accessed, we can store the matrix in column-wise straight storage format. These storage formats allow the accessing of diagonal elements also simultaneously. Neither of these techniques is good for an algorithm requiring access to both row and column vectors. The skewed storage format allows such access. But the diagonal elements are not accessible simultaneously in the skewed storage format.

In general, a conflict-free access occurs if in a modular (parallel) memory system with $M$ memory modules, $M$ consecutive accesses to the elements of matrix (vector) are spread over $M$ memory modules. In an interleaved memory system connected to processors through an $n$-by-$n$ switch, a necessary condition is that the memory access delay must be at most equal to $M$ clock periods. If this condition is not satisfied, after $M$ accesses an access to the first module is made which is still busy since the memory access time is greater than $M$. This leads to a conflict.

---

**Example 5.7**
Consider the storage scheme shown below where a dummy column is added to the matrix so that the dummy element in each row provides a cyclical offset to the row. This way no two elements of any row or any column are stored in the same module and thus the access is conflict-free.

| M0 | M1 | M2 | M3 |
|----|----|----|----|
| a00 | a01 | a02 | a03 |
|     | a10 | a11 | a12 |
| a13 |     | a20 | a21 |
| a22 | a23 |     | a30 |
| a31 | a32 | a33 |     |

M0, M1, M2, M3 are the memory modules.

---

The address increment required to move from one element to another in vector access is known as *stride*. For the above data structure, the stride for row access is 1, that for column access is 5, and that for diagonal access is 6.

A general rule is that stride $s$ for an access to a vector (matrix) stored in modular memory should be relatively prime to $m$, the number of memory modules. That is,

$$\text{GCD}(s, m) = 1 \tag{5.2}$$

for $m$ successive accesses to be spread across $m$ modules. Here, GCD denotes the greatest common divisor. Thus the access of diagonal elements in the above example is not conflict free, since the stride is 6. Budnick and Kuck (1971) suggested using memory with a prime number of modules. The reasoning behind this suggestion was that if $M$ is prime then

$$\text{GCD}(s, m) = 1 \tag{5.3}$$

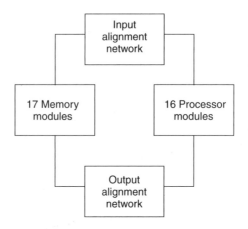

**FIGURE 5.9**
Burroughs Scientific Processor

for all $s < m$. This gives a wide range of choices for the stride, which satisfy the condition for conflict-free access.

This concept was used in the design of the BSP which had 17 memory modules, and 16 processors. The block diagram of the BSP is as shown in Figure 5.9. Alignment networks were used to twist the data to the required position, once accessed from the memory or before storing into memory.

---

**Example 5.8**
The following example shows the vector access in computers with prime number of memory modules:
Here $m$ is 5 and a (4 × 4) matrix with 2 dummy columns is shown.

| M0  | M1  | M2  | M3  | M4  |
|-----|-----|-----|-----|-----|
| a00 | a01 | a02 | a03 |     |
|     | a10 | a11 | a12 | a13 |
|     |     | a20 | a21 | a22 |
| a23 |     |     | a30 | a31 |
| a32 | a33 |     |     |     |

where M0 M1 M2 M3 M4 are Memory modules
Stride for row access = 1
Stride for column access = 6
Diagonal access with stride = 7

---

# Array Processors

|     | M0  | M1  | M2  | M3  | M4  |
| --- | --- | --- | --- | --- | --- |
|     | a00 | a01 | a02 | a03 |     |
|     |     | a10 | a11 | a12 | a13 |
|     |     |     | a20 | a21 | a22 |
|     | a23 |     |     | a30 | a31 |
|     | a32 | a33 |     |     |     |

|     |     |     |     |     |     |
| --- | --- | --- | --- | --- | --- |
| a00 | a33 | a11 |     | a22 | READ |
| a00 | a11 | a22 | a33 |     | ALIGN |

Some of the problems with BSP and other architectures using a prime number of memory modules are

1. Delay due to input and output alignment networks.
2. Addressing is complex when $M$ is not a power of 2.
3. Because each access has to be from the main memory, buffering of data is not possible, and hence results in a large delay in terms of data access.

The skewed storage scheme characteristics can be generalized. The skewed storage of Figure 5.6 is usually denoted as (1,1) skewed storage since the stride in both dimensions of the matrix is 1.

Let $k$ be the array dimension, $m$ be the number of memory modules, and $s_i$ is the stride in $i$th dimension, where $1 \leq i \leq k$. For a two-dimensional array in general with a $(s1, s2)$-skewing scheme, the columns will be said to be $s1$-ordered, the rows $s2$-ordered, and the main diagonal $(s1 + s2)$-ordered.

Let a $d$-ordered $n$ vector (mod $m$) denote a vector of $n$ elements, the $i$th element of which is stored in memory module $(di + c)(\bmod m)$, where $c$ is an arbitrary constant. Then a sufficient condition for a conflict-free access to such a vector is (Budnick and Kuck, 1971)

$$m \geq n \times \mathrm{GCD}(d, m) \tag{5.4}$$

**Example 5.9**
If a two-dimensional array is stored in $(s1, s2)$ order in $m$ memory modules to obey the above rule, we can access without conflict: any

s1-ordered $n$-element column since $m > n \times \text{GCD}(s1, m)$; any s2-ordered $n$-element row since $m > n \times \text{GCD}(s2, m)$; the (s1 + s2)-ordered main diagonal since $m > n \times \text{GCD}(s1 + s2, m)$.

Note that, it must be possible to index into each of the memory modules independently, in order to exploit skewing schemes effectively.

## 5.3 Interconnection Networks

The PEs and memory blocks in an SIMD architecture are interconnected by either an $n$-to-$n$ or an $n$-by-$n$ switch as discussed earlier. Also, an $n$-by-$n$ switch between the PEs is needed in general to allow fast exchange of data as the computation proceeds. Here onward, the term *interconnection network (IN)* will be used to refer to the interconnection hardware rather than the term "switch." The term "switch" is more generally used to refer to a component of an IN (as discussed later in this section).

As mentioned earlier, the PE-to-PE interconnection network for an SIMD depends on the data flow requirements of the application. In fact, depending on the application most of the computation can be accomplished by IN itself if the network is chosen appropriately, thus reducing the complexity of PEs drastically.

Several INs have been proposed and built over the last few years. This section introduces the terminology and performance measures associated with INs and describes common INs, as applied to SIMD systems. Chapter 6 extends this description to MIMD architectures.

### 5.3.1 Terminology and Performance Measures

An IN consists of several *nodes* interconnected by *links*. A node in general is either a PE or a memory block or a complete computer system consisting of PEs, memory blocks, and I/O devices. A link is the hardware interconnect between two nodes. The IN facilitates the transmission of *messages* (data or control information) between processes residing in nodes. The two functional entities that form the interconnection structure are *paths* and *switches*. A path is the medium by which a message is transmitted between two nodes and comprises one or more links and switches. The link just transmits the message and does not alter it in any way. A switch on the other hand, may alter the message (i.e., change the destination address) or route it to one of the number of alternative paths available.

# Array Processors

A path can be unidirectional point-to-point, bidirectional point-to-point, or bidirectional and visit more than two nodes. The first two types are classified as *dedicated* paths and the last type is the *shared* path.

Two message transfer strategies are used: *direct* and *indirect*. In the direct strategy, there will be no intervening switching elements between communicating nodes. In the indirect strategy, there will be one or more switching elements between the nodes. If an indirect transfer strategy is chosen, either a *centralized* or a *decentralized* transfer control strategy can be adopted. In a central strategy, all switching is done by a single entity (called the switch controller). In a decentralized strategy on the other hand, the switch control is distributed among a number of switching elements.

For instance, in a ring network there is a path from each node to every other node. The path to the neighboring node from any node is of length 1, while the path between non-neighboring nodes is of length equal to the number of links that need to be traversed (i.e., number of hops needed) to transmit messages between those nodes. If a decentralized control strategy is used, each intermediate node in the path serves as a switch that decodes the destination address and transmits the message not addressed to it to the next node. In the examples provided earlier in this chapter, the CP was the controller of the ring and issues multiple shift instructions depending on the distance (see below) between the source and destination nodes.

In general, an IN should be able to connect all the nodes in the system to one another, transfer maximum number of messages per second reliably, and offer minimum cost. Various performance measures have been used to evaluate INs. They are described below:

*Connectivity* (or *degree* of the node) is the number of nodes that are immediate neighbors of a node, that is, the number of nodes that can be reached from the node in one hop. For instance, in a unidirectional ring each node is of degree 1 since it is connected to only one neighboring node. In a bidirectional ring each node is of degree 2.

*Bandwidth* is the total number of messages the network can deliver in unit time. A message is simply a bit pattern of certain length consisting of data and control information.

*Latency* is a measure of the overhead involved in transmitting a message over the network from the source node to the destination node. It can be defined as the time required to transmit a zero-length message.

*Average distance* is the "distance" between two nodes is the number of links in the shortest path between those nodes, in the network. The average distance is given by:

$$d_{\text{avg}} = \frac{\sum_{d=1}^{r} d \cdot N_d}{N - 1} \qquad (5.5)$$

where $N_d$ is the number of nodes at distance $d$ apart, $r$ is the *diameter* (i.e., the maximum of the minimum distance between all pairs of nodes) of the network, and $N$ is the total number of nodes. It is desirable to have a low average distance. A network with low average distance would result in nodes of higher degree, that is, larger number of communication ports from each node, which may be expensive to implement. Thus, a normalized average distance can be defined as

$$d_{\text{avg(normal)}} = d_{\text{avg}} \times P \qquad (5.6)$$

where $P$ is the number of communication ports per node.

*Hardware complexity* of the network is proportional to the total number of links and switches in the network. As such, it is desirable to minimize the number of these elements.

*Cost* is usually measured as the network hardware cost as a fraction of the total system hardware cost. The "incremental hardware cost" (i.e., *Cost modularity*) in terms of how much additional hardware and redesign is needed to expand the network to include additional nodes is also important.

*Place modularity* is a measure of expandability, in terms of how easily the network structure can be expanded by utilizing additional modules.

*Regularity*: If there is a regular structure to the network, the same pattern can be repeated to form a larger network. This property is especially useful in the implementation of the network using VLSI circuits.

*Reliability and fault tolerance* is a measure of the redundancy in the network to allow the communication to continue, in case of the failure of one or more links.

*Additional functionality* is a measure of other functions (such as computations, message combining, arbitration, etc.) offered by the network in addition to the standard message transmission function.

A complete IN (i.e., an *n*-by-*n* network in which there is a link from each node to every other node — see Figure 5.13) is the ideal network since it would satisfy the minimum latency, minimum average distance, maximum bandwidth, and simple routing criteria. But, the complexity of this network becomes prohibitively large, as the number of nodes increases. Expandability also comes at a very high cost. As such, the complete interconnection scheme is not used in networks with large number of nodes. There are other topologies that provide a better cost/performance ratio.

The following design choices were used by Feng (1981) to classify INs (see Figure 5.10):

1. Switching mode
2. Control strategy

# Array Processors

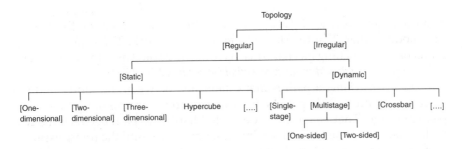

**FIGURE 5.10**
Feng's taxonomy (Feng, 1981)

3. Topology
4. Mode of operation

Switching modes and control strategies were defined earlier in this section. The IN topologies used in computer systems today have been adopted from the multitude of topologies used in telephone switching networks over the years. These topologies can be classified as either *regular* or *irregular*. Irregular topologies are usually a result of interconnection of existing isolated systems to allow communication between them. INs used in tightly coupled multiple-processor systems tend to be regular, in the sense that a definite pattern exists in the topology.

The topologies can further be classified as either *static* or *dynamic*. In a *static* topology, the links between two nodes are passive, dedicated paths. They cannot be changed to establish a direct connection to other nodes. The interconnections are usually derived based on a complete analysis of the communication patterns of the application. The topology does not change as the computation progresses. In *dynamic* topologies, the interconnection pattern changes as the computation progresses. This changing pattern is brought about by setting the network's active elements (i.e., switches).

The mode of operation of a network can be *synchronous*, *asynchronous*, or *combined* as dictated by the data manipulation characteristics of the application. The ring network used in the examples earlier in this chapter is an example of a synchronous network. In these networks communication paths are established and message transfer occurs synchronously. In asynchronous networks, connection requests are issued dynamically as the transmission of the message progresses. That means, the message flow in an asynchronous network is less orderly compared to that in a synchronous network. A combined network exhibits both modes of operation.

## 5.3.2 Routing Protocols

Routing is the mechanism that establishes the path between two nodes for transmission of messages. It should be as simple as possible and should not result in a high overhead. It should preferably be dependent on the state of each node, rather than the state of the whole network. Three basic routing protocols (or switching mechanisms) have been adopted over the years: *circuit switching, packet switching,* and *wormhole switching.*

In *circuit switching,* a path is first established between the source and the destination nodes. This path is dedicated for the transmission of the complete message. That is, there is a dedicated hardware path between the two nodes that cannot be utilized for any other message until the transmission is complete. This mode is ideal when large messages are to be transmitted.

In *packet switching,* the message is broken into small units called *packets.* Each packet has a destination address that is used to route the packet to the destination through the network nodes, in a *store and forward* manner. That is, a packet travels from the source to the destination, one link at a time. At each (intermediate) node, the packet is usually stored (or buffered). It is then forwarded to the appropriate link based on its destination address. Thus, each packet might follow a different route to the destination and the packets may arrive at the destination in any order. The packets are reassembled at the destination to form the complete message. Packet switching is efficient for short messages and more frequent transmissions. It increases the hardware complexity of the switches, because of the buffering requirements.

The packet switching mode is analogous to the working of the postal system in which each letter is a packet. Unlike messages in an IN, the letters from a source to a destination follow the same route and they arrive at the destination usually in the same order they are sent. The circuit switching mode is analogous to the telephone network in which a dedicated path is first established and maintained throughout the conversation. The underlying hardware in the telephone network uses packet switching. But what the user sees is a (virtual) dedicated connection.

The *wormhole switching* (or *cut-through routing*) is a combination of the above two methods. Here, a message is broken into small units (called *flow control digits* or *flits*), as in packet switching. But, all flits follow the same route to the destination, unlike packets. Since the leading flit sets the switches in the path and the others follow, the store and forward buffering overhead is reduced.

Routing mechanisms can be either *static* (or *deterministic*) or *dynamic* (or *adaptive*). Static schemes determine a unique path for the message from the source to the destination, based solely on the topology. Since they do not take into consideration the state of the network, there is the potential of uneven usage of the network resources and congestion. The dynamic routing on the other hand, utilizes the state of the network in determining the path for a message and hence can avoid congestion by routing messages around the heavily used links or nodes.

The routing is realized by setting switching elements in the network appropriately. As mentioned earlier, the switch setting (or control) function can either be managed by a *centralized controller* or can be *distributed* to each switching element. Further detail on routing mechanisms is provided in the following sections, along with the description of various topologies.

### 5.3.3 Static Topologies

Figures 5.11 to 5.14 show some of the many static topologies that have been proposed, representing a wide range of connectivity. The simplest among these is the linear array in which the first and the last nodes are each connected to one neighboring node and each interior node is connected to two neighboring nodes. The most complex is the complete IN in which each node is connected to every other node. A brief description of each of these topologies follows.

#### 5.3.3.1 Linear array and ring (loop)

In the one-dimensional mesh, or linear array [Figure 5.11(a)], each node is connected to its neighboring node, each interior node is connected to two of its neighboring nodes while the boundary nodes are connected to only one neighbor. The links can either be unidirectional or bidirectional.

A ring or loop network is formed by connecting the two boundary nodes of the linear array [Figure 5.11(b)]. In this structure, each node is connected to two neighboring nodes. The loop can either be unidirectional or bidirectional. In a unidirectional loop, each node has a source neighbor and a destination neighbor. The node receives messages only from the source and sends messages only to the destination neighbor. Messages circulate around the loop from the source with intermediate nodes acting as buffers to the destination. Messages can be of fixed or variable length and the loop can be designed such that either one or multiple messages can be circulating simultaneously.

The logical complexity of the loop network is very low. Each node should be capable of originating a message destined for a single destination, recognize a message destined for itself, and relay messages not destined for it. Addition of a node to the network requires only one additional link and flow of messages is not significantly affected by this additional link. The fault tolerance of the loop is very low. The failure of one link in a unidirectional loop causes the communication to stop (at least between the nodes connected by that link). If the loop is bidirectional, the failure of one link does not break the loop. Two link failure partitions the loop into two disconnected parts. Loops with redundant paths have been designed to provide the fault

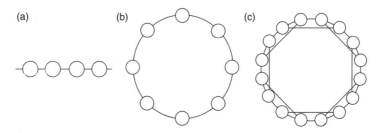

**FIGURE 5.11**
One-dimensional networks: (a) linear array, (b) ring, and (c) chordal ring

tolerance. The chordal ring (Figure 5.11(c)) in which the degree of each node is 3, is an example. The bandwidth of the loop can be a bottleneck as the communication requirements increase. It is also possible that one node can saturate the entire bandwidth of the loop.

Loop network evolved from the data communication environments in which geographically dispersed nodes were connected for file transfers and resource sharing. They have used bit-serial data links as the communication paths.

Note that when a loop network is used to interconnect PEs in an SIMD system, the message transfer between PEs is simultaneous since the PEs work in a lock-step mode. The transfer is typically controlled by the CP through SHIFT or ROTATE instructions issued once (for transfer between neighboring PEs) or multiple times (for transfers between remote PEs). The links typically carry data in parallel, rather than in bit-serial fashion.

### 5.3.3.2  Two-dimensional mesh

A popular IN is the two-dimensional mesh (*nearest-neighbor*) shown in Figure 5.12(a). Here the nodes are arranged in a two-dimensional matrix form and each node is connected to four of its neighbors (north, south, east, and west).

The connectivity of boundary nodes depends on the application. The IBM wire routing machine (WRM) uses a "pure mesh" in which boundary nodes have degree 3, and corners have degree 2. In the mesh network of ILLIAC-IV the bottom node in a column are connected to the top node in the same column and the rightmost node in a row is connected to the leftmost node in the next row. This network is called a *torus*.

The cost of two-dimensional network is proportional to $N$, the number of nodes. The network latency is $N$. The message transmission delay depends on the distance of the destination node to the source and hence there is a wide range of delays. The maximum delay increases with $N$.

A higher-dimensional mesh can be constructed analogous to the one- and two-dimensional meshes above. A $k$-dimensional mesh is constructed by arranging its $N$ nodes in the form of a $k$-dimensional array and connecting

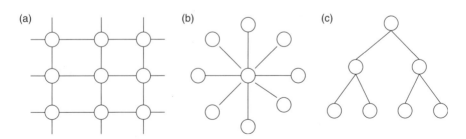

**FIGURE 5.12**
Two-dimensional networks: (a) near-neighbor mesh, (b) star, and (c) tree

each node to its ($2^k$) neighbors by dedicated, bidirectional links. The diameter of such a network is $k\sqrt[k]{N}$.

The routing algorithm commonly employed in mesh networks follows a traversal of one dimension at a time. For instance, in a four-dimensional mesh, to find a path from the node labeled (a, b, c, d) to the node labeled (p, q, r, s), we first traverse along the first dimension to the node (p, b, c, d), then along the next dimension to (p, q, c, d), then along the third dimension to (p, q, r, d), and finally along the remaining dimension to (p, q, r, s).

It is not possible to add a single node to the structure in general. The number of nodes and the additional hardware needed depends on the mesh dimensions. As such, the cost- and place-modularities are poor. Because of the regular structure, the routing is simpler. The fault tolerance can be high, since the failure of a node or a link can be compensated by other nodes and other alternative paths possible.

### 5.3.3.3   Star

In the star interconnection scheme [Figure 5.12(b)], each node is connected to a central switch through a bidirectional link. The switch forms the apparent source and destination for all messages. The switch maintains the physical state of the system and routes messages appropriately.

The routing algorithm is trivial. The source node directs the message to the central switch which in turn directs it to the appropriate destination on one of the dedicated links. Thus, if either of the nodes involved in message transmission is the central switch, the message path is just the link connecting them, if not, the path consists of two links.

The cost- and place-modularity of this scheme are good with respect to PEs, but poor with respect to the switch. The major problems are the switch bottleneck and the catastrophic effect on the system in the case of the switch failure. Each additional node added to the system requires a bidirectional link to the switch and the extension of switch facilities to accommodate the new node.

Note that the central switch basically interconnects the nodes in the network. That is, the interconnection hardware is centralized at the switch, often allowing centralized instead of distributed control over routing. The interconnection structure within the switch could be of any topology.

### 5.3.3.4   Binary trees

In binary tree networks [Figure 5.12(c)], each interior node has a degree 3 (two children and one parent), leaves have degree 1 (parent), and the root has degree 2 (two children).

A simple routing algorithm can be used in tree networks. In order to reach node Y from node X, traverse up in the tree from X until an ancestor of Y is reached and then traverse down to Y. To find the shortest path, we need to first find the ancestor of Y at the lowest level in the tree while ascending from X and then decide whether to follow right or left link while descending towards Y.

The nodes of the tree are typically numbered consecutively, starting with the root node at 1. The node numbers of the left and right children of a node $z$ are $2z$ and $2z + 1$, respectively. Further, the node numbers at level $i$ (with the root level being 1) will be $i$ bits long. Thus the numbers for the left and right children of a node can be obtained by appending a 0 (left child) or a 1 (right child) to the parent's number. With this numbering scheme, in order to find a path from node X and to node Y, we first extract the longest common bit pattern from the numbers of X and Y. This is the node number of their common ancestor A. The difference in the lengths of numbers of X and A is the number of levels to be traversed up from X. In order to reach Y from A, we first remove the common (most significant) bits in numbers A and Y. We then traverse down based on the remaining bits of Y, going left on 0 and right on 1, until all the bits are exhausted (Almasi and Gottlieb, 1989).

For a tree network with $N$ nodes, the latency is $\log_2 N$, the cost is proportional to $N$, the degree of nodes is 3 and is independent of $N$.

### 5.3.3.5 *Complete interconnection*

Figure 5.13 shows the complete IN, in which each node is connected to every other node with a dedicated path. Thus, in a network with $N$ nodes, all nodes are of degree $N - 1$ and there are $N(N - 1)/2$ links. Since all nodes are connected, the minimal length path between any two nodes is simply the link connecting them.

Routing algorithm is trivial. The source node selects the path to the destination node among the $(N - 1)$ alternative paths available and all nodes must be equipped to receive messages on a multiplicity of paths.

This network has poor cost modularity, since the addition of a node to an $N$ node network requires $N$ extra links and all nodes must be equipped with additional ports to receive the message from the new node. Thus, the complexity of the network grows very fast as the number of nodes is increased. Hence, complete INs is used in environments where only a small number of nodes (4–16) are interconnected. The place modularity is also poor for the same reasons as for the cost modularity.

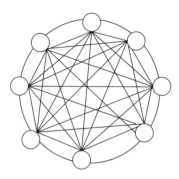

**FIGURE 5.13**
Completely connected network

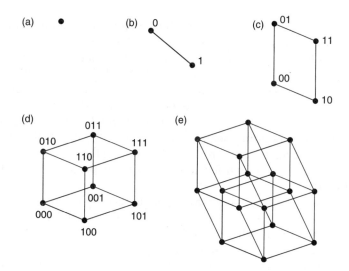

**FIGURE 5.14**
Hypercubes of various values of $k$: (a) zero-cube, (b) one-cube, (c) two-cube, (d) three-cube, and (e) four-cube

This network provides a very high bandwidth and its fault tolerance characteristics are good. Failure of a link does not make the network inoperable, since alternative paths are readily available (although the routing scheme gets more complex).

The Intel Paragon, the MIT Alewife (Agarwal et al., 1995), and the MPP SIMD computer (Batcher, 1980) use two-dimensional networks. A three-dimensional mesh network is used in the J-machine at MIT (Dally et al., 1993). The MasPar MP-1 (Blank, 1990; Nickolls, 1992) and MP-2 (MasPar, 1992) use a two-dimensional torus network where each node is connected to its eight nearest neighbors by the use of shared X connections. The Fujitsu AP3000 distributed-memory multicomputer has a two-dimensional torus network (Ishihata et al., 1997), while the CRAY T3D (Kessler and Schwarzmeier, 1993; Koeninger et al., 1994) and T3E both use the three-dimensional torus topology.

### 5.3.3.6  Hypercube

The hypercube is a multidimensional near-neighbor network. A $k$-dimensional hypercube (or a $k$-cube) contains $2^k$ nodes, each of degree $k$. The label for each node is a binary number with $k$ bits. The labels of neighboring nodes differ in only one bit position.

Figure 5.14 shows the hypercubes for various values of $k$. A 0-cube has only one node as shown in Figure 5.14(a). A 1-cube connects 2 nodes labeled 0 and 1 as shown in Figure 5.14(b). A 2-cube of Figure 5.14(c) connects four nodes labeled: 00, 01, 10, and 11. Each node is of degree 2 and the labels of neighboring nodes differ in only one bit position. A 3-cube as shown in Figure 5.14(d) has 8 nodes with labels ranging from 000 through 111 (or decimal 0 through 7). Node (000) for example is connected directly to nodes (001), (010), and (100) and the message transmission between these neighboring nodes requires only one hop. To transmit a message from node (000) to node (011), two routes are possible: 000 to 001 to 011 and 000 to 010 to 011. Both routes require two hops. Note that the source and destination labels differ in two bits, implying the need for two hops. To generate these routes a simple strategy is used. First the message is at 000. The label 000 is compared with the destination address 011. Since they differ in bit positions 2 and 3, the message is routed to one of the corresponding neighboring nodes 010 or 001. If the message is at 010, this label is again compared with 011, noting the difference in position 3, implying that it be forwarded to 011. A similar process is used to route from 001 to 011.

Thus, the routing algorithm for the hypercube is simple. For a $k$-dimensional hypercube the routing algorithm uses at most $k$ steps. During step $i$ the messages are routed to the adjacent node in dimension $i$ if the $i$th bit of X is 1; otherwise, the messages remain where they are.

Hypercube networks reduce network latency by increasing the degree of each node (i.e., connecting each of the $N$ nodes to $\log_2 N$ neighbors). The cost is of the order of $N \log_2 N$ and the latency is $\log_2 N$.

Several commercial architectures using the hypercube topology (from Intel Corporation, NCUBE Corporation, and Thinking Machines Corporation) are now available. One major disadvantage of the hypercube topology is that the number of nodes should always be a power of two. Thus, the number of nodes need to double, every time a single node is required to be added to the network.

### 5.3.4 Dynamic Topologies

A parallel computer system with static interconnection topology can be expected to do well on applications that can be partitioned into processes with predictable communication patterns consisting mostly of exchanges among neighboring processing elements. Some examples of such application domains are the analysis of events in space, vision, image processing, weather modeling, and very large-scale integration (VLSI) design. If the application does not exhibit such predictable communication pattern, machines with static topologies become inefficient since the message transmission between non-neighboring nodes results in excessive transmission delays. Hence, computer systems using static INs tend to be more special-purpose compared to those using dynamic INs. Both SIMD and MIMD architectures have used static topologies.

Several dynamic INs have been proposed and built over the last few years, with a wide range of performance/cost characteristics. They can be classified under the following categories:

1. Bus networks
2. Crossbar networks
3. Switching networks

### 5.3.4.1 Bus networks

These are simple to build. Several standard bus configurations have evolved with data path widths as high as 64 bits. A bus network provides the least cost among the three types of dynamic networks and also has the lowest performance. Bus networks are not suitable for PE interconnection in an SIMD system. Chapter 6 provides further details on bus networks in the context of MIMD architectures.

### 5.3.4.2 Crossbar networks

A crossbar network is the highest performance, highest cost alternative among the three dynamic network types. It allows any PE to be connected to any other non-busy PE at any time.

Figure 5.15 shows an $N \times N$ crossbar, connecting $N$ PEs to $N$ memory elements. The number of PEs and memory elements need not be equal, although both are usually powers of 2. The number of memory elements is usually a small multiple of the number of PEs. There are $N^2$ crosspoints in the crossbar, one at each row–column intersection. If the PEs produce a 16-bit address and work with 16-bit data units, each crosspoint in the crossbar corresponds to the intersection of 32 lines plus some control lines. Assuming 4 control lines, to build a $16 \times 16$ crossbar we would need at least ($16 \times 16 \times 36$) switching devices. To add one more PE to the crossbar, only one extra row of crosspoints is needed. Thus, although the wire cost grows as the number of processors $N$, the switch cost grows as $N^2$, which is the major disadvantage of this network.

In the crossbar network of Figure 5.15, any PE can be connected to any memory and each of the $N$ PEs can be connected to a distinct memory, simultaneously. In order to establish a connection between a PE and a memory block the switch settings at only one crosspoint need to be changed. Since there is

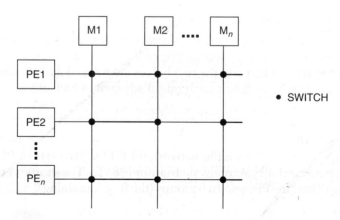

**FIGURE 5.15**
$N \times N$ crossbar

only one set of switches in any path, the crossbar offers uniform latency. If two PEs try to access the same memory, then there is a contention and one of them needs to wait. Such contention problems can be minimized by memory organizations discussed in Section 5.2.

The operation of the MIMD requires that the processor–memory interconnections be changed in a dynamic fashion. With high-speed switching needed for such operation, high frequency capacitive and inductive effects result in noise problems that dominate crossbar design. Because of the noise problems and high cost, large crossbar networks are not practical.

The complete connectivity offered by crossbars may not be needed always. Depending on the application, a "sparse" crossbar network in which only certain crosspoints have switches may be sufficient, as long as it satisfies the bandwidth and connectivity requirements.

The above description uses memory blocks as one set of nodes connected to a set of PEs. In general, each node could either be a PE, or a memory block or a complete computer system with PE, memory and I/O components.

In an SIMD architecture using crossbar IN for PEs, the switch settings are changed according to the connectivity requirements of the application at hand. This is done either by the CP or a dedicated switch controller. Alternatively, a decentralized control strategy can be used where the switches at each cross-point forward the message toward the destination node.

### 5.3.4.3 Switching networks

*Single-* and *multi-stage* switching networks offer a cost/performance compromise between the two extremes of bus and crossbar networks. A majority of switching networks proposed are based on an interconnection scheme known as perfect shuffle. The following paragraphs illustrate the single- and multi-stage network concepts as applied to SIMD architectures, through the perfect shuffle IN. The discussion of other switching networks is deferred to Chapter 6.

### 5.3.4.4 Perfect shuffle (Stone, 1971)

This network derives its name from its property of rearranging the data elements in the order of the perfect shuffle of a deck of cards. That is, the card deck is cut exactly in half and the two halves are merged such that the cards in similar position in each half are brought adjacent to each other.

---

**Example 5.10**

Figure 5.16 shows the shuffle network for 8 PEs. Two sets of PEs are shown here for clarity. Actually there are only 8 PEs. The shuffle network first partitions the PEs into two groups (the first containing 0, 1, 2, and 3

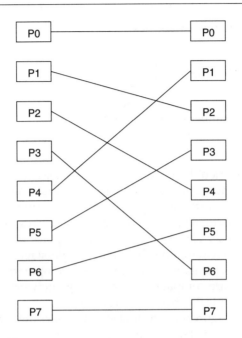

**FIGURE 5.16**
Perfect shuffle

and the second containing 4, 5, 6, and 7). The two groups are then merged such that 0 is adjacent to 4, 1 to 5, 2 to 6 and 3 to 7.

The interconnection algorithm can also be derived from a cyclic shift of the addresses of PEs. That is, number PEs starting from 0, with each PE number consisting of $n$ bits where the total number of PEs is $2^n$. Then, to determine the destination of PE after a shuffle, shift the $n$-bit address of the PE left once cyclically. The shuffle shown in Figure 5.16 is thus derived as the following:

| PE | Source | Destination | PE |
|----|--------|-------------|----|
| 0 | 000 | 000 | 0 |
| 1 | 001 | 010 | 2 |
| 2 | 010 | 100 | 4 |
| 3 | 011 | 110 | 6 |
| 4 | 100 | 001 | 1 |
| 5 | 101 | 011 | 3 |
| 6 | 110 | 101 | 5 |
| 7 | 111 | 111 | 7 |

> This transformation can be described by
>
> $$\text{Shuffle}(i) = \begin{cases} 2i & \text{if } i < N/2 \\ 2i - N + 1 & \text{if } i \geq N/2 \end{cases} \quad (5.7)$$
>
> where $N$ is the number of PEs and is a power of 2.

The shuffle network of Figure 5.16 is a *single-stage* network. If an operation requires multiple shuffles to complete, the network is used multiple times. That is, the data *recirculate* through the network. Figure 5.17 shows a *multi-stage* shuffle network for 8 PEs in which each stage corresponds to one shuffle. That is, the data inserted into the first stage ripples through the multiple stages rather than recirculating through a single stage. In general, multi-stage network implementations provide faster computations at the expense of increased hardware compared to single-stage network implementations. The following example illustrates this further.

In the network of Figure 5.17, the first shuffle makes the vector elements that were originally $2^{(n-1)}$ distance apart adjacent to each other; the next shuffle brings the elements originally $2^{(n-2)}$ distance apart adjacent to each other, and so on. In general, the $i$th shuffle brings the elements that were $2^{(n-i)}$ distance apart adjacent to each other. In addition to the shuffle network, if the PEs are

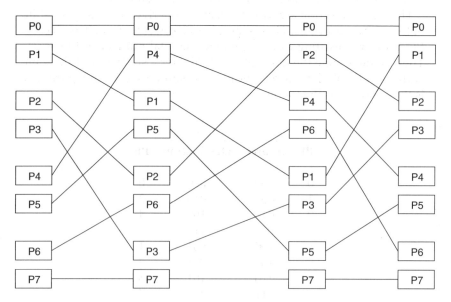

**FIGURE 5.17**
Movement of data in a perfect shuffle network

# Array Processors

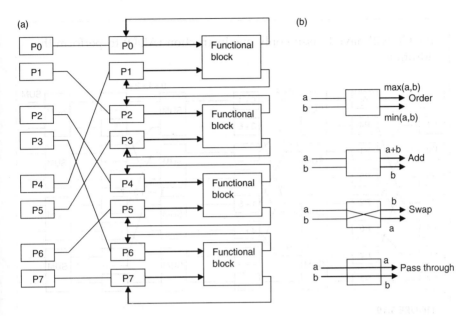

**FIGURE 5.18**
Processors connected by perfect shuffle: (a) network and (b) representative functions

connected such that adjacent PEs can exchange data, the combined network can be used efficiently in several computations.

Figure 5.18 shows a network of 8 PEs. In addition to the perfect shuffle, adjacent PEs are connected through a function block capable of performing several operations on the data from the PEs connected to it. Several possible functions are shown in Figure 5.18(b). The add function returns the sum of the two data values to the top PE. The order function rearranges data in increasing or decreasing order. The swap function exchanges the data values and the pass through function retains them as they are.

### Example 5.11
Figure 5.19 shows the utility of the network of Figure 5.18 in the accumulation of $N$ data values in the PEs. Each stage in the network shuffles the data and adds the neighboring elements with the sum appearing on the upper output of the SUM block. All the four SUM blocks are required in the first stage. Only the first and the third SUM blocks are needed in the second stage and only the first one in the last stage, for this computation. If each shuffle-and-add cycle consumes $T$ time units, this addition requires only $\log_2 N \times T$ time units compared to $N \times T$ time units required by a sequential process. Also note

the CP will have to issue only one instruction (ADD) to perform this addition.

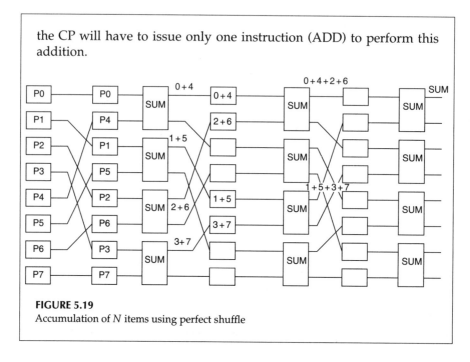

**FIGURE 5.19**
Accumulation of $N$ items using perfect shuffle

The addition of $N$ elements can also be accomplished by the single-stage network of Figure 5.18, where each functional block will be a SUM block. Now the data has to circulate through the network $\log_2 N$ times. Thus the CP will have to execute the following program rather than a single instruction (as in the implementation with multi-stage network):

```
for i = 1 to log_2 N
        shuffle
        add
endfor
```

Assuming $T$ time units for shuffle and add as before, the single-stage network implementation requires an execution time of: $T \times \log_2 N + x$ Fetch/decode time for instructions in the loop body each time through the loop + the loop control overhead.

### Example 5.12
Figure 5.20 shows the sorting of $N$ data values using the shuffle-operate network above. Here, the operate function corresponds to the functional block rearranging the data elements in adjacent PEs in decreasing order (if required). Again, sorting requires order of $\log_2 N$ time.

# Array Processors

| Processor | Data | Shuffle | Operate | Shuffle | Operate | Shuffle | Operate |
|---|---|---|---|---|---|---|---|
| 0 | 5 | 5 | 5 | 5 | 9 | 9 | 9 |
| 1 | 6 | 3 | 3 | 9 | 5 | 7 | 7 |
| 2 | 9 | 6 | 6 | 3 | 3 | 5 | 6 |
| 3 | 4 | 2 | 2 | 1 | 1 | 6 | 5 |
| 4 | 3 | 9 | 9 | 6 | 7 | 3 | 4 |
| 5 | 2 | 1 | 1 | 7 | 6 | 4 | 3 |
| 6 | 1 | 4 | 7 | 2 | 4 | 1 | 2 |
| 7 | 7 | 7 | 4 | 4 | 2 | 2 | 1 |

       Cycle 1      Cycle 2      Cycle 3

**FIGURE 5.20**
Sorting using perfect shuffle

For the above sort operation it is required that the original data be *bitonic*. A vector data is bitonic if (a) the vector elements increase monotonically and then decrease monotonically (with either the increasing or decreasing portions to be empty) or (b) a cyclic shift of the vector elements satisfy the property (1).

There are several switching networks based on the perfect shuffle concept. For further details on INs refer to Chapter 6, the review article by Feng (1981) and books by Hillis (1985), and Siegel (1985).

## 5.4 Performance Evaluation and Scalability

If the algorithm and data partitioning allow a degree of parallelism $N$, an SIMD with $N$ processors can execute the algorithm in a single time unit, that is, $O(1)$. This is an enhancement of execution speed by $N$-fold (i.e., the speedup is $N$), since the execution time of the same algorithm on an SISD would be of the $O(N)$. If the SIMD has only $p$ PEs where $p < N$, the data are partitioned into $N/p$ segments and the PEs operate simultaneously on the $p$ elements of each segment. The computation then would be of the $O(N/p)$ and hence the speedup is $p$.

The speedup $S$ provided by a parallel computer system is usually defined as:

$$S = \frac{\text{Sequential execution time}}{\text{Parallel execution time}} \quad (5.8)$$

In general, there can be more than one sequential algorithm to solve a given problem. On an SISD, it is natural to use the fastest among the available algorithms. It is possible that either the fastest algorithm is not known or if known, it may be very difficult to implement it on the given SISD system. Then we would use the fastest among the known algorithms that is implementable. The numerator of equation (5.8) corresponds to the running (or execution) time of such an algorithm.

All the sequential algorithms available to solve a problem may not be equally suitable for parallel implementation. In particular, the best sequential algorithm may not be the one that is most parallelizable. Thus, the denominator of the above equation is the running time of the most efficient parallel algorithm for the given architecture.

Theoretically, the maximum speedup possible with a $p$ processor system is $p$. It is possible to obtain a speedup greater than $p$ (i.e., a *superlinear* speedup), if either a nonoptimal sequential algorithm has been used in the computation of sequential run time or some hardware characteristic results in excessive execution time for the sequential algorithm. Consider for instance, an application to compute the sum of a large set ($N$) of numbers. If the complete dataset does not fit into the main memory of the SISD, the overall computation time would be increased since some of the data has to be retrieved from the secondary memory. When the same algorithm is implemented on an SIMD with $p$ processors, the data are partitioned into $p$ sets of $N/p$ units each. Each of these sets may be small enough to fit in the main memory. Thus, the overall execution time would be less than $N/p$, resulting in a speedup of greater than $p$.

---

**Example 5.13**
Consider the problem of accumulating $N$ numbers. The execution time on an SISD is of the $O(N)$. On an SIMD with $N$ processors and a ring interconnection network between the processors, the execution consists of $(N-1)$ communication steps and $(N-1)$ additions, resulting in a total time (assuming communication and addition each take 1 time unit) of $2(N-1)$ or $O(2N)$. Thus, the speedup is 0.5.

---

If the processors in the SIMD are interconnected by a perfect shuffle network (as in Figure 5.19), the execution consists of $\log_2 N$ shifts and $\log_2 N$ additions, resulting in a total time of $2\log_2 N$ or $O(\log_2 N)$. Hence the speedup is of the $O(N/\log_2 N)$.

# Array Processors

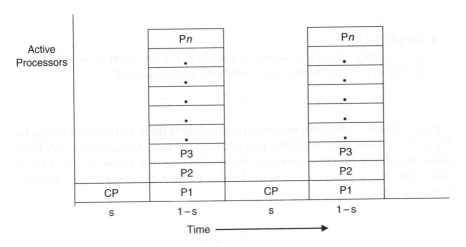

**FIGURE 5.21**
Computation cycle in an array processor

Ideally, an SIMD system with $p$ processors should deliver a speedup of $p$. This is not possible in practice, since all the processors in the system cannot be kept busy performing useful computations all the time, due to the characteristics of the algorithms. The timing diagram of Figure 5.21 illustrates the operation of a typical SIMD system. Here, the horizontal axis represents time and the processor activity is plotted on the vertical axis. For a short period of time only the CP is busy initializing variables, fetching and decoding instructions to be sent to the arithmetic processors, and performing other serial computations. Following this serial period, the arithmetic processors perform the computation in parallel. This mode of operation repeats with each parallel segment preceded by a serial segment. The *efficiency* is a measure of the fraction of time that the processors are busy. In Figure 5.21, the fraction of time spent in serial code is $s$. That is, only one processor is busy during this time. The fraction of time spent in parallel code is $(1-s)$, when all the $p$ processors are busy. If the number of processors is $p$, the efficiency $E$ is given by:

$$E = s/p + (1-s)$$
$$= 1 - s(1 - 1/p) \tag{5.9}$$

As $p$ increases, the efficiency tends to $(1-s)$, the serial code execution time remains $s$, but the parallel code execution time shortens (assuming that the degree of parallelism offered by the algorithms is at least $p$). Thus, the serial code dictates the efficiency. In an ideal system, $s$ is 0 and hence $E$ is 1. In practice, $E$ is between 0 and 1 depending on the degree to which the processors are utilized.

> **Example 5.14**
> The efficiency of accumulating $N$ numbers on an $N$-processor SIMD with perfect shuffle network is of the order of $(1/\log_2 N)$.

If we normalize the computation times such that the serial execution time in Figure 5.21 is one unit, and if the code that can be run in parallel takes $N$ time units on a single-processor system, the parallel execution time shrinks to $N/p$ when run on a $p$ processor system. Then, $s$ is the ratio of serial time to total execution time:

$$s = \frac{1}{1 + N/p}$$
$$= \frac{p}{p + N} \tag{5.10}$$

For large values of $p$, $s$ tends to 1 and hence the efficiency approaches 0, since the parallelizable portions of code take vanishingly small fraction of time while the serial code occupying only one processor makes the remaining processors idle.

The efficiency is also defined as the ratio of speedup to the number of processors in the parallel system. Thus,

$$E = \frac{S}{p} \tag{5.11}$$

The *cost* of solving a problem on a parallel architecture is defined as the sum of the time that each of the $p$ processors is busy solving that problem. The cost is thus the product of the parallel run time and the number of processors $p$ (i.e., *processor-time product*). A parallel architecture is said to be *cost-optimal*, if the cost of solving a problem on that architecture is proportional to the execution time of the fastest algorithm for solving that problem on a single-processor system.

The *scalability* of a parallel system is a measure of its ability to increase speedup as the number of processors increases. A parallel system is *scalable*, if its efficiency can be maintained at a fixed value, by increasing the number of processors as the *problem size* increases. Here, the problem size is defined as the total number of operations required to solve the problem. For instance, the problem size of computing the sum of an array of $N$ elements is $N$; the problem size of addition of two $N \times N$ matrices is $N^2$; and that of multiplication of two $N \times N$ matrices is $N^3$.

In general, the SIMD systems are inherently scalable, since the processor and interconnection hardware can simply be replicated as the problem size increases.

It is ideal to execute an algorithm with a degree of parallelism $N$ on an SIMD with $N$ processors, thus obtaining the maximum possible speedup. As the problem size increases, the hardware complexity and cost usually dictate that the number of processors cannot be arbitrarily increased. Then, the parallel system is *scaled down* in terms of the number of processors. That is, an algorithm designed to run on an $N$ processor system (i.e., $N$ *virtual processors*) is mapped to run on $p$ *physical processors*, where $N > p$. Thus, each of the $p$ physical processors emulates $N/p$ virtual processors. In general, there will be multiple mappings of virtual to physical processors. For the computation to be efficient, the mapping should be such that it minimizes the communication time between the virtual processors by mapping them to near-neighbor physical processors as far as possible.

---

**Example 5.15**
The problem of accumulating $N$ numbers can be solved in two methods on an SIMD with $p$ processors and a perfect shuffle interconnection network. Here, $p < N$ and we assume that $N/p$ is less than or equal to $p$. In the first method, each block of $p$ numbers are accumulated in $O(\log_2 p)$. Since there are $N/p$ such blocks, the execution time is $O(N/p \times \log_2 p)$. The resulting $N/p$ partial sums are then accumulated in $O(\log_2 p)$. Thus, the total run time is $O(N/p \times \log_2 p + \log_2 p)$. In the second method, each of the $p$ blocks of $N/p$ numbers is allocated to a processor. The run time for computing the partial sums is then $O(N/p)$. These partial sums are accumulated using the perfect shuffle network in $O(\log_2 p)$. Thus, the total run time is $O(N/p + \log_2 p)$. The second method offers a better run time than the first.

If $N/p$ is greater than $p$, then further portioning of the computations to fit the $p$ processor structure is needed. Computation of run time for this case is left as an exercise.

---

The performance of an SIMD thus depends not only on the algorithm employed, but also on the data partitioning techniques and the interconnection structure used.

In addition to the above theoretical performance measures, the comparison of high performance computers is usually done based on their peak computation rates in terms of MFLOPS and MIPS. This mode of comparison does not reflect the true operating environment since the peak performance rate of an array processor can only be obtained under the most ideal circumstances. Of much greater validity are the sustained performance rates. Sustained performance is usually measured by benchmarks that reflect accurately the types of workloads expected on these processors.

## 5.5 Programming SIMDs

As shown earlier in this chapter, the instruction set of an SIMD system is similar to that of an SISD, except that the arithmetic/logic instructions are executed by all the PEs simultaneously and the control instructions are executed by the CP. The SIMD instruction set contains additional instructions for IN operations, manipulating local and global registers and setting activity bits based on data conditions. Programming SIMDs at the assembly language level is very similar to that with SISDs.

A high-level programming language for SIMD systems, should allow architecture independent programming. For instance, the programmer programs assuming an appropriate number of virtual processors based on the problem size, rather than the actual number of (physical) processors in the system. The compiler then maps the virtual processors to physical processors and generates the code for communication between the processors. Thus, the compiler complexity will be higher than that for traditional SISD compilers. The language should also allow the explicit representation of the data parallelism.

Popular high-level programming languages such as FORTRAN, C, and Lisp have been extended to allow data-parallel programming on SIMDs. Sections 5.6 and 5.7 provide further details on these language extensions. There are also compilers that translate serial programs into data-parallel object codes.

An algorithm that is efficient for SISD implementation may not be efficient for an SIMD, as illustrated by the matrix multiplication algorithm of Section 5.3. Thus, the major challenge in programming SIMDs is in devising an efficient algorithm and corresponding data partitioning such that all the PEs in the system are kept busy throughout the execution of the application. This also requires minimizing conditional branch operations in the algorithm.

The data exchange characteristics of the algorithm dictate the type of IN needed. If the desired type of IN is not available, routing strategies that minimize the number of hops needed to transmit data between non-neighboring PEs will have to be devised.

## 5.6 Example Systems

The Intel and MIPS processors described in Chapter 1 have SIMD features. The supercomputer systems described in Chapter 4 also operate in an SIMD mode. This section provides brief descriptions of the hardware, software, and application characteristics of two SIMD systems. The ILLIAC-IV has been the most famous experimental SIMD architecture and is selected for its historical interest. Thinking Machine Corporation's Connection Machine series, although no longer in production, originally envisioned for data-parallel symbolic computations, and later allowed numeric applications.

### 5.6.1 ILLIAC-IV

The ILLIAC-IV project was started in 1966 at the University of Illinois. The objective was to build a parallel machine capable of executing $10^9$ instructions per second. To achieve this speed, a system with 256 processors controlled by a control processor was envisioned. The set of processors was divided into 4 quadrants of 64 processors each, each quadrant to be controlled by one control unit. Only one quadrant was built and it achieved a speed of $2 \times 10^8$ instructions per second.

Figure 5.22 shows the system structure. The system is controlled by a Burroughs B-6500 processor. This machine compiles the ILLIAC-IV programs, schedules array programs, controls array configurations, and manages the disk file transfers and peripherals. The disk file unit acts as the backup memory for the system.

Figure 5.23 shows the configuration of a quadrant. The control unit (CU) provides the control signals for all processing elements ($PE_0$–$PE_{63}$), which work in an instruction lock-step mode. The CU executes all program control instructions and transfers processing instructions to PEs. The CU and the PE-array execute in parallel. In addition, the CU generates and broadcasts the addresses of operands that are common to all PEs, receives the status signals from PEs, from the internal I/O operations, and from B-6500 and performs the appropriate control function.

Each PE has four 64-bit registers (accumulator, operand register, data routing register, and general storage register), an unit arithmetic logic unit (ALU),

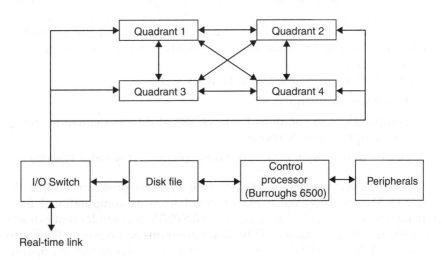

**FIGURE 5.22**
ILLIAC-IV structure

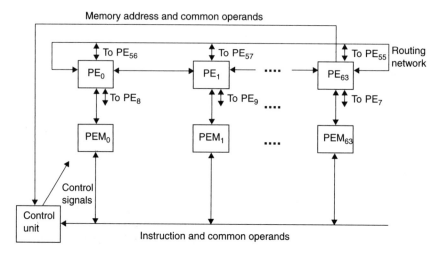

**FIGURE 5.23**
A quadrant of ILLIAC-IV

a 16-bit local index register, and an 8-bit mode register that stores the processor status and provides the PE enable-disable information. Each processing element memory (PEM) block consists of a 250 nanosecond cycle-time memory with 2K, 64-bit words.

The PE-to-PE routing network connects each PE to four of its neighbors. The PE array is arranged as an 8 × 8 torus. Interprocessor data communication of arbitrary distances is accomplished by a sequence of routings over the routing network.

The processing array is basically designed to operate on 64-bit operands. But, it can be configured to perform as a 128, 32-bit subprocessor array or a 512, 8-bit subprocessor array. The subprocessor arrays are not completely independent because of the common index register in each PE and the 64-bit data routing path.

The applications envisioned for ILLIAC-IV were

1. Manipulation of large matrices.

2. Computations of solutions for large sets of different equations for weather-prediction purposes.

3. Fast data correlation for fast-response, phased-array radar systems.

Two high-level languages (HLLs) were used to program ILLIAC-IV: a FORTRAN-like language (CFD) oriented towards computational fluid dynamics and an ALGOL-like language, GLYPNIR. A compiler that extracts parallelism from conventional FORTRAN programs and converts them into parallel FORTRAN (IVTRAN) was also developed. There was no operating system. Being an experimental machine, the ILLIAC-IV did not offer either a software library or any software development tools, thereby making it difficult to program. The ILLIAC-IV is now located at the NASA Ames Center and is not operational.

### 5.6.2 Thinking Machine Corporation's CM-2

The CM-2 introduced in 1987, is a massively parallel SIMD machine configured with up to 64K single-bit processors, providing a performance in the range of 2.5 GFLOPS. Table 5.1 summarizes its characteristics.

The CM-2 hardware consists of one to four front-end computers, a parallel processing unit (PPU) and an I/O system that supports mass storage and graphic display devices (Figure 5.24). It can be viewed as a data-parallel extension to the conventional front-end computers since control flow and serial operations are handled by the front-end computers while the operations that can be run efficiently in parallel are executed on the PPU. A 4 × 4 crossbar switch (Nexus) connects the front-end computers to the four sections of the PPU. Each section is controlled by a sequencer which decodes the assembly language (Paris) instructions and generates a series of micro instructions for each processor. The sequencers are implemented using Advanced Micro Devices' 2901/2910 bit-sliced processor with 16K, 96-bit words of

**TABLE 5.1**

CM characteristics

| | | |
|---|---|---|
| General | Processors | 65.536 |
| | Memory | 512 MB |
| | Memory bandwidth | 300 Gbits sec$^{-1}$ |
| I/O channels | Number of channels | 8 |
| | Capacity per I/O controller | 40 MB sec$^{-1}$ |
| | Total I/O controller transfer rate | 300 MB sec$^{-1}$ |
| | Capacity per framebuffer | 1 Gbits sec$^{-1}$ |
| Typical application performance (fixed point) | General computing | 2500 MIPS |
| | Terrain mapping | 1000 MIPS |
| | Document search | 6000 MIPS |
| Interprocessor communication | Regular pattern of 32-bit messages | 250 million sec$^{-1}$ |
| | Random pattern 32-bit messages | 80 million sec$^{-1}$ |
| | Sort 64K 32-bit keys | 30 million sec$^{-1}$ |
| Variable precision fixed point | 64-bit integer add | 1500 MIPS |
| | 16-bit integer add | 3300 MIPS |
| | 64-bit move | 2000 MIPS |
| | 16-bit move | 3800 MIPS |
| Double-precision floating point | 4K × 4K matrix multiply benchmark | 2500 MIPS |
| | Dot product | 5000 MIPS |
| Single precision floating point | Addition | 4000 MFLOPS |
| | Subtraction | 4000 MFLOPS |
| | Multiplication | 4000 MFLOPS |
| | Division | 1500 MFLOPS |
| | Peak performance | 32 GFLOPS |

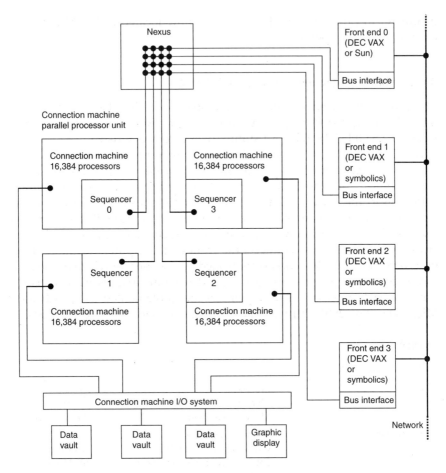

**FIGURE 5.24**
Architecture of CM-2 (Courtesy of Thinking Machine Corporation.)

microcode memory. Interprocessor communication is handled by either a hypercube interconnection network or a faster multidimensional grid that allows processors to simultaneously transmit data in regular patterns.

### 5.6.2.1 Processor

The processors are implemented using four different chip types: the processor chip, the memory chip, and two floating-point accelerator chips. The processor chip (Figure 5.25) contains the ALU, flag registers, NEWS interface, router interface, and I/O interface for 16 processors. The 16 processors are connected by a 4 × 4 mesh allowing processors to communicate with their North, South, East, and West (NEWS) neighbors. The 64K bits of bit-addressable memory is implemented using commercial RAM chips. First of the floating-point chips contains an interface which passes the data on to the second chip which is the floating-point execution unit. These two chips are

## Array Processors

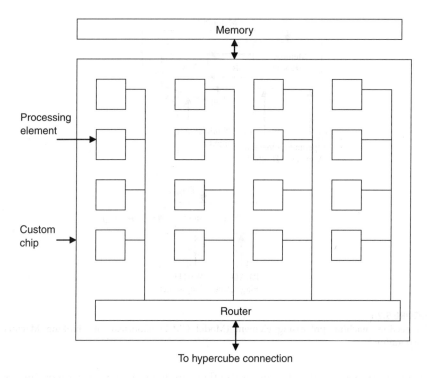

**FIGURE 5.25**
Connection machine custom chip (Courtesy of Thinking Machines Corporation.)

required for every 32 processors when floating-point acceleration is needed. Thus, each section of the PPU contains 1024 custom processor chips, 512 floating-point interface chips, 512 floating-point execution chips, and 128 MB of RAM.

The 16 flag registers include 8 bits for general purpose use and 8 bits that have functions predefined by the hardware. Special purpose flags include the NEWS flag which can be written to directly from the ALU of one of the adjacent processing elements, two flags, for message router data movement and handshaking, a memory parity flag, a flag for daisy chaining processing elements, a zero flag for reading or for operations that do not need to write to a flag, and two flags used primarily for diagnostics.

The CM-2 processor cell executes only one simple but powerful instruction that produces outputs based on a memory/flag lookup table. Figure 5.26 shows a processing cell for CM-1, however, the basic structure is the same for CM-2 (with memory increased from 4K to 64K). The ALU executes this instruction by taking three 1-bit inputs and producing two 1-bit outputs. Two of the inputs are taken from that processor's memory and the third bit is taken from one of the flag bits. The outputs are then calculated from 1 of 256 possible functions found in the memory/flag lookup table using two 8-bit control bytes provided by the sequencer. One output bit is written back into the memory

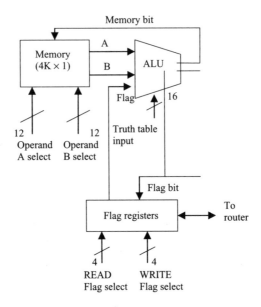

**FIGURE 5.26**
Connection machine processing element (Model CM-1) (Courtesy of Thinking Machines Corporation.)

and the other is written to one of the flag bits for subsequent processing. All processors receive the same instruction from the sequencer, however, individual processors may be masked off using a flag bit that controls processor activation.

The ALU cycle is broken up into subcycles. On each cycle the data processors can execute one low-level instruction (called a nanoinstruction) from the sequencer and the memories can perform one read or write operation. For instance, the basic ALU cycle for a two-operand integer add consists of three nanoinstructions:

LOAD A: read memory operand A, read flag operand, latch one truth table.

LOAD B: read memory operand B, read condition flag, latch other truth table.

STORE: store memory operand A, store result flag.

Arithmetic is performed in a bit serial fashion. At about a half a microsecond per bit plus overhead, a 32 bit add takes about 21 $\mu$sec. With the maximum of 64K processors all computing in parallel this produces a rate of around 2500 MIPS.

The floating-point accelerator performs both single and double precision operations on data in IEEE standard floating-point format, increasing the rate of floating-point operations by more than a factor of 20.

# Array Processors

### 5.6.2.2 Hypercube

The processors are linked by a 12-dimensional hypercube router network. Message passing occurs in a data-parallel mode, that is, all processors can simultaneously send data into the local memories of other processors, or fetch data from the local memories of other processors into their own. The communication hardware supports certain message-combining operations. For instance, the processors to which multiple messages are sent receive the bitwise logical OR from the ALU output of all the messages, or the numerically largest, or the integer sum.

The simplest mode of communication is broadcasting a value from the front-end system to all the PEs. A context flag within each PE controls, which PEs receive the broadcast value. When large amounts of data must be loaded into the Communication Machine memory, the I/O mechanism is generally faster. Regular, grid-based communication is used with data arranged in an $n$-dimensional grid or torus.

The following parallel communications operations permit elements of parallel variables to be rearranged and combined into parallel variables of different geometries:

1. *Reduce and Broadcast*: applies an operator (such as Sum) across the elements of a parallel variable.
2. *Grid (NEWS)*: shifts the elements of a parallel variable by a fixed amount along each dimension of the variable.
3. *General (Send, Get)*: processors send or get values from arbitrary processor locations. Provision is made for simultaneous access and combining of colliding data.
4. *Scan*: applies a binary associative operator to a sequence of processors. Cumulative results are returned in a new parallel variable. Scans implement parallel prefix operations over a set of data elements and form an important new class of parallel computation and communication operations. Scans may also be performed over a sequence of processors partitioned into disjoint sets. This provides the algorithm designer with a mechanism for simultaneously processing data arranged into multiple, variable-length sets.
5. *Spread*: takes one or more data elements and spreads them throughout a processor set. In the case of an $n$-dimensional grid, a spread may be used to copy an $n - 1$ subspace along a single dimension.
6. *Sort*: sorts elements in ascending or descending order of key.

### 5.6.2.3 Nexus

The Nexus is a $4 \times 4$ cross-point switch that connects up to 4 front-end computers to up to 4 sequencers. The connections to front-end computers are through high-speed, 32-bit, parallel, asynchronous data paths, while the connections to sequencers are 32-bit synchronous paths. The CM can be configured up to four sections under front-end control, using the Nexus. These sections can be for different users or purposes. Any front-end can be connected to any section or valid combination of sections. For instance, in a CM-2 system with 32K processors (i.e., four 8K processor sections) and four front-ends, all four sections can be ganged to be controlled by any one front-end; two 8K sections can be assigned to each of the two front-ends while the other two sections are ganged to provide 16K processors to a third front-end and the fourth front-end is used for purposes not related to the PPU.

### 5.6.2.4 Router

The routing network is used to transmit data from one processor chip to the other. Messages travel from one router node to another until they reach the chip containing the destination processor. Suppose a processor $j$ on chip $i$ wants to communicate with processor $l$ on chip $k$. It first sends the message to the router on its own chip. This router forwards the message to the router on chip $k$, which in turn delivers the message to the appropriate memory location for processor $l$. The routing algorithm used by the router moves messages across each of the 12 dimensions of the hypercube in sequence. If there are no conflicts, a message will reach its destination within one cycle of this sequence, since any vertex of the cube can be reached from any other by traversing no more than 12 edges. The router allows every processor to send a message to any other processor, with all messages being sent and delivered at the same time. The messages may be of any length. The throughput of the router depends on the message length and on the pattern of accesses.

The entities that operate on data in parallel within the CM are the physical processors. However, each physical processor can simulate any number of conceptual processing entities called virtual processors. This simulation is transparent to the user. The router logic supports the 'virtual processor' concept. When a message is to be delivered by a router node, it is placed not only within the correct physical processor, but in the correct region of memory for the virtual processor originally specified as the message's destination.

### 5.6.2.5 NEWS grid

The CM-1 NEWS grid is a two-dimensional mesh that allows nearest-neighbor communication. In the CM-2, the two-dimensional grid was replaced by a more general $n$-dimensional grid. This allows nearest-neighbor grid communication of multiple dimensions. Possible grid configurations for 64K processors are: $256 \times 256$, $1024 \times 64$, $8 \times 8192$, $64 \times 32 \times 32$, $16 \times 16 \times 16 \times 16$, and $8 \times 8 \times 4 \times 8 \times 8 \times 4$. The NEWS grid allows processors to pass data according to a regular rectangular pattern. The advantage of this mechanism over the router is that the overhead of explicitly specifying destination addresses is eliminated.

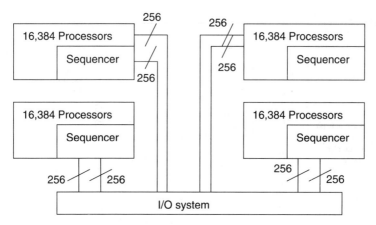

**FIGURE 5.27**
Connection machine I/O system (Courtesy of Thinking Machine Corporation.)

### 5.6.2.6 Input/output system

Each 8K-processor section is connected to one of the eight I/O channels (Figure 5.27). Each I/O channel may be connected to either a high resolution graphics display framebuffer module or an I/O controller. I/O transfers are initiated by the front-end computers through Paris instructions to the sequencers, which cause direct, parallel transfers between the I/O devices and the data processors.

Each I/O channel consists of 256 data bits and is switched through sequencer control among two 4K banks of processors (i.e., 256 chips × 16 processors = 4K processors and 256 parallel bits of data). Data is passed along the channels (see Figure 5.28) to the I/O controllers with parity checked for each byte and stored in one of 512, 288-bit buffers (32 parity bits + 256 data bits). When the buffers are full, the I/O controller signals that it is ready to send data to a Data Vault.

Each I/O controller is connected to one or more Data Vaults via an 80-bit bus. Data Vaults combine very high reliability with very fast transfer rates for large blocks of data. With a capacity of 10 GB each, eight Data Vaults operating in parallel can achieve transfer rates of 40 MB sec$^{-1}$. Data Vaults consist of 39 individual disk drives and 3 spare drives in case of failures. The 64 data bits received from the I/O controller are split into two 32-bit words and 7 Error Correcting Code (ECC) bits are added to each. The 39 bits are stored in 39 disk buffers which are eventually written to individual drives. The ECC permits 100% recovery of data if one drive fails.

As an alternative to the I/O controller, a high resolution graphics display framebuffer module may be connected to one of the eight I/O channels. This module supports a high-speed, real-time, tightly coupled graphical display, utilizing a 1280 × 1024-pixel frame buffer module with 24-bit color and four overlays.

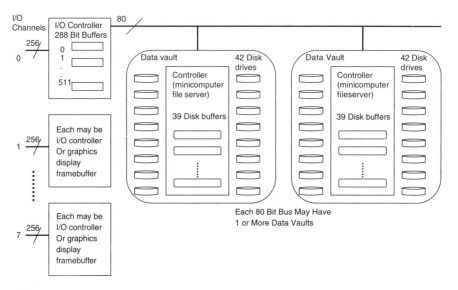

**FIGURE 5.28**
Connection machine data vault (Courtesy of Thinking Machine Corporation.)

### 5.6.2.7 Software

System software is based upon the operating system or environment of the front-end computers, with minimal software extensions. Thus, users can program using familiar languages and program constructs, with all development tools provided by the front-end. The flow of control is handled entirely by the front-end, including storage and execution of the program and all interaction with the user and programmer. In addition to the assembly language Paris (Parallel instruction set), the HLLs used for programming the CM are *Lisp, CM-Lisp, and C*. These are data-parallel dialects of Common Lisp and C, respectively.

### 5.6.2.8 Paris

Paris includes operations such as integer arithmetic, floating-point arithmetic, inter processor communication, vector summation, matrix multiplication, and sorting. The front-end sends Paris instructions to the PPU. The program control structures are not part of the Paris instruction set since they are handled by the front-end computer.

The Paris user interface consists of a set of functions, subroutines, and global variables. The functions and subroutines direct the actions of the CM processors, and the variables allow the user program to find out such information about the CM system as the number of processors available and the amount of memory per processor.

Consider the following C code:

```
while (CM_f_gt_constant(z, 1.0, 23, 8),
       CM_global_logior(CM_test_flag, 1)) {
CM_f_divide_constant_2(z, 2.0, 23, 8);
}
```

It repeatedly causes every processor whose floating-point z field is greater than 1.0 to be divided by two; the loop is terminated when no processor has a z value greater than one. Here, the functions whose names begin with "CM_" are Paris operations. CM_f_gt_constant causes every processor to compare a field to a common, broadcast constant, storing a bit reflecting the result in its "test" flag. CM_f_divide_constant similarly causes every processor to divide a floating-point field by a common constant; and CM_global_logior takes a bit field (test flag) from every processor, and returns to the front-end as a result of a many-way bitwise inclusive-OR operation. The while construct is not part of the Paris language.

### 5.6.2.9  *Lisp

The *Lisp is an extension of Common Lisp for programming the CM in a data-parallel style. The parallel primitives of *Lisp support the CM model. That is, each processor executes a subset of Common Lisp, with a single thread of control residing on the front-end. The language supports all of the standard Common Lisp data types, in addition to a parallel data type called pvar(parallel variable).

The Common Lisp feature of run-time type checking is retained and hence *Lisp also requires no declarations. The standard functions for reading or writing the contents of a variable (pref and pref-grid) in combination with the macro setf are used to store data from the front-end into a pvar of a processor. The assignment operator (*set) stores the value of a pvar expression into the destination pvar. Each of the following parallel functions accept a scalar and return a pvar that contains the scalar in all active processors:

```
mod!!  ash!!  round!!  integerp!!

max!!  min!!  if!!  eql!!

ldb!!  dpb!!  byte!!  numberp!!
```

### 5.6.2.10 C*

C* is an extension C designed to support data-parallel programming. It features a new data type based on classes in C++, a synchronous execution model, and some extensions to C syntax. Pointers are used for interprocessor communication. A synchronous computation model is assumed by C*, where all instructions are issued from the front-end. It allows different processors of CM to have different memory layouts, since they may hold different kinds of data. A structure type called domain allows the specification of the memory layout. Accordingly, the C* code is divided into serial and parallel portions. The code that belongs to a domain is parallel and is executed by many data processors, while the serial code is executed by the front-end.

The data is divided into scalar and parallel portions, described by keywords, mono and poly, respectively. Mono data resides in the memory of the front-end, while the poly data resides in the memory of the data processors. There are no new operators for parallel computation. At compile-time distinction is made between the two types of data and scalar operators are extended (through overloading) to operate on parallel data.

There is one new statement type in C*, the selection statement. This is used to active multiple processors. The C* compiler parses the C* source code, performs type and data flow analysis, and translates parallel code into a series of function calls that invoke CM Paris operations.

### 5.6.2.11 *Applications*

Paris presents the user with an abstract machine architecture very much like the physical CM hardware but with two important extensions: a much richer instruction set and the virtual processor abstraction. It provides a set of parallel primitives ranging from simple arithmetic and logical operations to high level, APL-like reduction (parallel prefix) operations, sorting, and communications. The virtual processor mechanism supported at the paris level is transparent to the user. When the system is initialized, the number of processors required by the application is specified. If this number exceeds the available physical processors the local memory of each processor splits into as many regions as necessary with the processors automatically time sliced among the regions.

The CM has been used in the following application areas:

1. Document retrieval from large databases, analysis of English text, and memory based reasoning
2. Circuit simulation and optimization of cell placement
3. Object recognition and image processing
4. Computer-generated graphics for animation and scientific visualization
5. Neural net simulation, conceptual clustering, classifier systems, and genetic algorithms
6. Modeling geological strata using reverse time migration techniques

7. Protein sequence matching
8. $N$-body interactions, such as modeling defect movement in crystals under stress and modeling galaxy collisions
9. Fluid flow modeling, cellular autonoma, and navier stokes based simulation

A brief description of document retrieval application on the CM follows. Locating information in natural language documents contained in large databases is an important problem. Two general methods of accessing such databases are: free text search and key word search. Free text search systems allow the user to search the text of the documents for arbitrary combinations of words, whereas keyword systems require that indexers read all the documents in a database and assign them keywords from a controlled vocabulary. In the latter system, the user searches for information or documents using some combination of keywords from this controlled vocabulary.

The CM employs the "relevance search" technique. The user provides a list of keywords or a question. These words are broadcast to all the processors along with the numerical weights associated with each word. Each processor notes whether each word occurs in it's article; if so, corresponding numerical score is incremented, based on a weighting scheme. The weight assigned to a word is inversely proportional to its frequency in the database. This weighting mechanism ensures that uncommon words have higher influence than the common words. When all the words have been compared, the article with the largest scores are retrieved and presented to the user. The user can browse through the documents returned and find one or more documents of interest. In the document database, one article is stored per processor, in a compressed form (called content kernel). The source document is stored in the front-end. Each word of the content kernel is represented as a bit vector and stored in the form of a table called kernel. This coding method is called surrogate encoding, in which each word is mapped into $n$ different bits by using $n$ hash functions. On each keyword, the processor applies the $n$ hash functions to calculate the hash number. This number is then compared with each entry of the kernel and if a match occurs the score of that article is incremented. The articles with highest scores are then retrieved.

## 5.7 Summary

Architectural issues concerned with parallel array processor or SIMD schemes are covered in this chapter. These architectures are suitable for computational environments that exhibit a very high degree of data-parallelism. They can be considered as a special subset of MIMD architectures that offer the least amount of overhead. This chapter also provided a description of hardware and software characteristics of two SIMD systems: Thinking Machine Corporation's Connection Machine and an experimental SIMD (ILLIAC-IV).

In general, an efficient algorithm for an SISD system will not be efficient on an SIMD. New algorithms and correspondingly specialized programming languages are needed to efficiently utilize the computational power offered by SIMD architectures. Data structuring and memory organization also contribute heavily to the efficiency of SIMD systems.

The processor interconnection network is a major component of an SIMD. These networks are tailored to the application at hand and hence make the architecture specialized to the application. Common interconnection networks were introduced in this chapter. Chapter 6 will extend this description as applied to MIMD architectures. Interconnection network development is an active area of research (Merritt, 2004).

## Problems

5.1. The following are common operations on matrices: column sum, row sum, transpose, inverse, addition, and multiplication. Examine the algorithm for each operation and develop vectorized procedures for each assuming an array processor capable of performing add, subtract, reciprocal, and multiply operations.

5.2. Assume $N \times N$ matrix for each operation in the above problem. Develop procedures suitable for an SIMD with $N$ processing elements. Specify how the matrices are represented in the $N$ memory blocks of the SIMD.

5.3. Solve Problem 5.2 for an SIMD with $M$ processing elements where $M < N$ and $N$ is not a multiple of $M$.

5.4. Develop procedures and data structures for the matrix operations in Problem (5.1) for execution on one quadrant of the ILLIAC-IV.

5.5. Consider the $8 \times 8$ mesh of ILLIAC-IV. An application requires that each PE update its content by the average of values in its four neighboring PEs. Write an assembly language program to run this computation for 5 iterations assuming:

1. Each PE computes the new value based on current contents of the neighbors and then updates its value.
2. The update is performed one row at a time starting from the top row of PEs.

5.6. Develop the architectural facilities needed, if the procedures in the above Problem need to run until a convergence condition is satisfied (the new value is different from the old value by a small $e$ in all the PEs, for instance).

5.7. Consider an 8 × 8 crossbar that interconnects 8 PEs. Show the active crosspoints needed in this crossbar if only the following interconnections are needed:

1. A unidirectional ring
2. A bidirectional ring (i.e., one-dimensional nearest neighbor)
3. Two-dimensional nearest-neighbor (as in ILLIAC-IV)
4. Three-dimensional hypercube
5. Shuffle exchange

5.8. Consider the nearest-neighbor mesh interconnection of the ILLIAC-IV with 64 PEs ($PE_0$–$PE_{63}$), organized as an 8 × 8 array. $PE_i$ is connected to $PE_{i+1}$, $PE_{i-1}$, $PE_{i+8}$, and $PE_{i-8}$. Determine the minimum number of routing steps needed to send data from $PE_i$ to $PE_j$, where $j = (i+k) \bmod 64$ and $0 \le k \le 63, 0 \le i \le 63$.

5.9. A plus-minus-$2i$ IN connects each $PE_j$ ($0 \le j \le N-1$) of an $N$ processor system to all $PE_{k+}$ and $PE_{k-}$, where $k+ = (j+2i) \bmod N$ and $k- = (j-2i) \bmod N$; and $0 \le i \le n, n = \log_2 N$.

1. Show the structure of the network for $N = 8$.
2. What is the maximum number of routing steps needed to transfer data from $PE_i$ to $PE_j$, where $i$ and $j$ are arbitrary integers and $0 \le j \le N-1, 0 \le i \le N-1$.
3. Show the single and multistage implementations.

5.10. It is required to broadcast a data value in one PE to all the other PEs in an SIMD system with $N$ PEs. Determine the number of steps needed if the network is each of the following types:

1. Single-stage shuffle-exchange
2. Hypercube

5.11. What is the maximum number of hops needed to route a single data item on:

1. A 64 × 64 processor array
2. A 4096-node hypercube?

5.12. Show that an $n$-cube has the same topology as an $n \times n$ array with toroidal edge connections.

5.13. List the desired characteristics of a compiler that performs parallelization of the sequential code for an SIMD.

5.14. Trace the evolution of the Connection Machine series (CM-1, CM-2, and CM-5), in terms of architectural characteristics and intended applications.

5.15. *Systolic array processors* and *Associative processors* are versions of SIMD architecture. Investigate the architectural differences.

# References

Agarwal, A., Bianchini, R., Chaiken, D., Johnson, K. L., Kranz, D., Kubiatowicz, J., Lim, B. H., Mackenzie, K., and Yeung, D. The MIT-Alewife machine: architecture and performance. In *Proceedings of the 22nd International Symposium on Computer Architecture*, June 1995.

Almasi, G. S. and Gottlieb, A. *Highly Parallel Computing*. Redwood City, CA: Benjamin Cummings, 1989.

Batcher, L. Design of a massively parallel processor. *IEEE Transactions on Computers*, 29, 1980, 836–840.

Blank, S. The MasPar MP-1 architecture. In *Proceedings of the 35th Annual IEEE International Computer Conference*, pp. 20–24, 1990.

Budnik, P. and Kuck, D. J. The organization and use of parallel memories. *IEEE Transactions on Computers*, C-20, 1971, 1566–1569.

Dally, J., Keen, S., and Noakes, M. D. The JMachine architecture and evaluation. In *Proceedings of the 38th Annual IEEE International Computer Conference*, pp. 183–188, 1993.

Feng, T.-Y. A survey of interconnection networks. *IEEE Computer*, 14, 1981, 12–27.

Hillis, W. D. *The Connection Machine*. Cambridge, MA: MIT Press, 1985.

Hillis, W. D. and Steele, G. L. *The Connection Machine Lisp Manual*. Cambridge, MA: Thinking Machine Corp., 1985.

Hord, R.M. *The ILLIAC-IV: The First Supercomputer*. Rockville, MD: Computer Science Press, 1983.

Ishihata, H., Takahashi, M., and Sato, H. Hardware of AP3000 scalar parallel server. *Fujitsu Scientific and Technical Journal*, 33, 1997, 24–30.

Kessler, R. E. and Schwarzmeier, J. L. CRAY T3D: A new dimension for cray research. In *Proceedings of the 38th Annual IEEE International Computer*, pp. 176–182, 1993.

Koeninger, R. K., Furtney, M., and Walker, M. A shared memory MPP from cray research. *Digital Technical Journal*, 6, 1994, 8–21.

MasPar Computer Corporation, The Design of the MasPar MP-2: A Cost Effective Massively Parallel Computer, Sunnyvale, CA, 1992.

Merritt, R. Choosing the right interconnect — welcome to the Giga Era. *Electronic Engineering Times*, 20/27, 2004, 31–38.

Nickolls, J. R. Interconnection architecture and packaging in massively parallel computers. In *Proceedings of the Packaging, Interconnects, and Optoelectronics for the Design of Parallel Computers Workshop*, March 1992, 161–170.

Shiva, S. G. *Computer Design and Architecture*. New York: Marcel Dekker, 2000.

Siegel, H. J. *Interconnection Networks for Large Scale Parallel Processing*. Lexington, MA: Lexington Books, 1985.

Skillicorn, D. B. A taxonomy for computer architectures. *IEEE Computer*, 21, 1985, 46–57.

Stone, H. S. Parallel processing with the perfect shuffle. *IEEE Transactions on Computers*, C-20, 1971, 153–161.

# 6
## Multiprocessor Systems

The most important characteristic of the multiprocessor systems discussed in this chapter is that all the processors function independently. That is, unlike the SIMD systems in which all the processors execute the same instruction at any given instant of time, each processor in a multiprocessor system can be executing a different instruction at any instant of time. For this reason, Flynn classified them as multiple instruction stream multiple data stream (MIMD) computers.

As mentioned earlier, the need for parallel execution arises since the device technology limits the speed of execution of any single processor. SIMD systems increased the performance and the speed manifold simply due to data-parallelism. But, such parallel processing improves performance only in a limited case of applications which can be organized into a series of repetitive operations on uniformly structured data. Since a number of applications cannot be represented in this manner, SIMD systems are not a panacea. This led to the evolution of a more general form, the MIMD architectures where in each processing element has its own arithmetic/logic unit (ALU) and control unit, and if necessary its own memory and input/output (I/O) devices. Thus each processing element is a computer system in itself, capable of performing a processing task totally independent of other processing elements. The processing elements are interconnected in some manner to allow exchange of data and programs and to synchronize their activities.

The major advantages of MIMD systems are

1. *Reliability:* If any processor fails, its workload can be taken over by another processor thus incorporating graceful degradation and better fault tolerance in the system.
2. *High performance:* Consider an ideal scenario where all the $N$ processors are working on some useful computation. At times of such peak performance, the processing speed of an MIMD is $N$ times that of a single processor system. However, such peak performance is difficult to achieve due to the overhead involved with MIMD operation.

The overhead is due to:

(a) Communication between processors
(b) Synchronization of the work of one processor with that of another processor
(c) Wastage of processor time if any processor runs out of tasks to do
(d) Processor scheduling (i.e., allocation of tasks to the processors).

A *task* is an entity to which a processor is assigned. That is, a task is a program, a function or a procedure in execution on a given processor. *Process* is simply another word for a task. A processor or a processing element is a hardware resource on which tasks are executed. A processor executes several tasks one after another. The sequence of tasks performed by a given processor in succession forms a *thread*. Thus the path of execution of a processor through a number of tasks is called a thread. Multiprocessors provide for simultaneous presence of a number of threads of execution in an application.

**Example 6.1**
Consider Figure 6.1 in which each block represents a task and each task has a unique number. As required by the application, task 2 can be executed only after task 1 is completed and task 3 can be executed only after task 1 and task 2 are both completed. Thus the line through tasks 1, 2, and 3 forms a thread (A) of execution. Two other threads (B and C) are also shown. These threads limit the execution of tasks to a specific

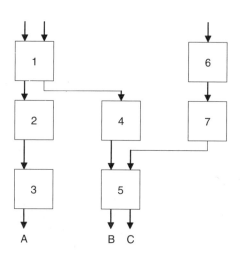

**FIGURE 6.1**
Threads of execution

serial manner. Although task 2 has to follow task 1 and task 3 has to follow task 2, note that task 4 can be executed in parallel with task 2 or task 3. Similarly, task 6 can be executed simultaneously with task 1, and so on. Suppose the MIMD has three processors and if each task takes the same amount of time to execute, the seven tasks shown in the figure can be executed in the following manner:

| Time Slot | Processor 1 | Processor 2 | Processor 3 |
|-----------|-------------|-------------|-------------|
| 1 | Task 1 | Task 6 | |
| 2 | Task 2 | Task 4 | Task 7 |
| 3 | Task 3 | | Task 5 |

Initially, since there are only two tasks that can be executed in parallel, they are allocated to processors 1 and 2 (arbitrarily) and processor 3 sits idle. At the end of the first time slot, all three processors get a task each from the three different threads. Thus tasks 2, 4, and 7 are executed in parallel. Finally tasks 3 and 5 are again executed in parallel with one of the processors sitting idle due to the lack of tasks at that instant.

Figure 6.1 implies that the application at hand is partitioned into seven tasks and exhibits a *degree of parallelism* of 3 since at the most three tasks can be executed simultaneously. It is assumed that there is no interaction between the tasks in this figure. In practice, tasks communicate with each other since each task depends on the results produced by other tasks. Refer to Figures 2.1, 2.2, and 2.3 for an illustration of the data dependencies between tasks. Obviously, the communication between tasks reduces to zero if all the tasks are combined into a single task and run on a single processor (i.e., SISD mode).

The $R/C$ ratio where $R$ is the length of the run time of the task and $C$ is the communication overhead produced by that task, signifies *task granularity*. This ratio is a measure of how much overhead is produced per unit of computation. A high $R/C$ ratio implies that the communication overhead is insignificant compared to computation time, while a low $R/C$ ratio implies that the communication overhead dominates computation time and hence a poorer performance.

High $R/C$ ratios imply *coarse-grain* parallelism while low $R/C$ ratios result in *fine-grain* parallelism. The general tendency to obtain maximum performance is to resort to the finest possible granularity, thus providing for the highest degree of parallelism. However, care should be taken to see that this maximum parallelism does not lead to maximum overhead. Thus a trade-off is required to reach an optimum level of granularity.

This chapter describes the architectural concepts, programming considerations, and applications of MIMD systems. The next section provides the models for the two common MIMD organizations. Section 6.2 describes the memory organization and the cache coherence problem. Section 6.3 extends the description of interconnection networks from Chapter 5 to MIMD architectures. Operating system considerations for multiprocessors are the theme of Section 6.4. Programming considerations are discussed in Section 6.5. Section 6.6 presents performance evaluation techniques and Section 6.7 provides a brief description of some commercial multiprocessor systems.

## 6.1 MIMD Organization

As mentioned earlier, in an MIMD system each processing element works independently of the others. Processor 1 is said to be working independently of processors $2, 3, \ldots, N$ at any instant, if and only if the task being executed by processor 1 has no interactions with the tasks executed by processors $2, 3, \ldots, N$ and vice versa at that instant. However, the results from the tasks executed on a processor X now may be needed by processor Y sometime in the future. In order to make this possible, each processor must have the capability to communicate the results of the task it performs to other tasks requiring them. This is done by sending the results directly to a requesting process or storing them in a shared-memory (i.e. a memory to which each processor has equal and easy access) area. These communication models have resulted in two popular MIMD organizations (presented in Chapter 2). These are

1. Shared Memory or tightly coupled architecture
2. Message Passing or loosely coupled architecture

### 6.1.1 Shared-Memory Architecture

Figure 6.2(a) shows the structure of a shared-memory MIMD. Here, any processor $i$ can access any memory module $j$ through the interconnection network. The results of the computations are stored in the memory by the processor that executed that task. If these results are required by any other task, they can be easily accessed from the memory. Note that each processor is a full-fledged SISD capable of fetching instructions from the memory and executing them on the data retrieved from the memory. No processor has a local memory of its own.

# Multiprocessor Systems

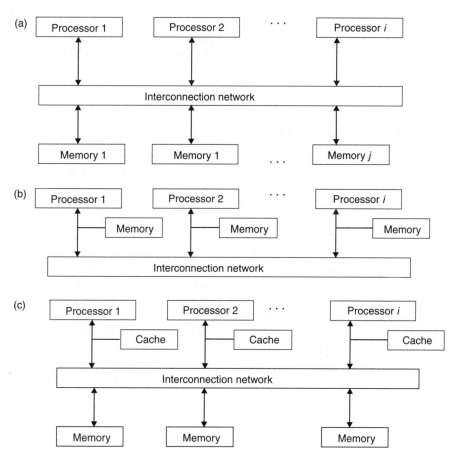

**FIGURE 6.2**
MIMD structures: (a) shared-memory system, (b) message-passing system, and (c) MIMD with processors having personal caches

This is called *tightly coupled* architecture, since the processors are interconnected such that the interchange of data between them through the shared memory is quite rapid. This is the main advantage of this architecture. Also, the memory access time is the same for all the processors and hence the name *uniform memory architecture* (UMA).

If the processors in the system are nonhomogeneous, data transformations will be needed during exchange. For instance, if the system consists of both 16-bit and 32-bit processors and the shared memory consists of 32-bit words, each memory word must be converted into two words for the use of 16-bit processors and vice-versa. This is an overhead.

Another problem is the memory contention, which occurs whenever two or more processors try to access the same shared-memory block. Since a memory block can be accessed only by one processor at a time, all the other processors requesting access to that memory block must wait until the first processor is through using it. Also if two processors simultaneously request the access to the same memory block, one of the processors should be given preference over the other. Memory organization concepts are discussed in Section 6.2.

### 6.1.2 Message-Passing Architecture

This is the other extreme, where there is no shared memory at all in the system. Each processor has a (local) memory block attached to it. The conglomeration of all local memories is the total memory that the system possesses. Figure 6.2(b) shows a block diagram of this configuration, also known as *loosely coupled* or *distributed memory* MIMD system. If data exchange is required between two processors in this configuration, the requesting processor $i$ sends a message to processor $j$, in whose local memory the required data are stored. In reply to this request the processor $j$ (as soon it can) reads the requested data from its local memory and passes it on to processor $i$ through the interconnection network. Thus, the communication between processors occurs through message passing.

The requested processor usually finishes its task at hand and then accesses its memory for the requested data and passes it on to the interconnection network. The interconnection network routes it toward the requesting processor. All this time the requesting processor sits idle waiting for the data, thus incurring a large overhead. The memory access time varies between the processors and hence these architectures are known as *nonuniform memory architectures* (NUMA). Thus, a tightly coupled MIMD offers more rapid data interchange between processors than a loosely coupled MIMD, while the memory contention problem is not present in a message-passing system since only one processor has access to a memory block.

### 6.1.3 Other Models

Shared-memory architectures are also known as *multiprocessor* systems, while message-passing architectures are called *Multicomputer* systems. These architectures are two extremes. MIMD systems in practice may have a reasonable mix of the two architectures as shown in Figure 6.2(c). In this structure, each processor operates in its local environment as far as possible. Inter-processor communication can be either by shared memory or by message passing.

Several variations of this memory architecture have been used. For instance, the data diffusion machine (DDM) (Hagersten et al., 1992) uses a cache only memory architecture (COMA) in which all system memory resides in large caches attached to the processors in order to reduce latency and network load. The IBM Research Parallel Processor (RP3) consists of 512 nodes, each containing 4 MB of memory. The interconnection of nodes is such that the 512 memory modules can be used as one global shared memory or purely as local memories with message-passing mode of communication or a combination of the both.

MIMD systems can also be conceptually modeled as either *private-address-space* or *shared-address-space* machines. Both address-space models can be implemented on shared memory and message-passing architectures. Private memory, shared-address-space machines are NUMA architectures that offer scalability benefits of message-passing architectures, with programming advantages of shared-memory architectures. An example of this type, is the J-machine from MIT, which has small private memory attached to each of a large number of nodes, but has a common address space across the whole system. The DASH machine from Stanford considers local memory as a cache for the large global address space, but the global memory is actually distributed. In general, the actual configuration of an MIMD system depends on the application characteristics for which the system has been designed.

## 6.2 Memory Organization

Two parameters of interest in MIMD memory system design are the *bandwidth* and the *latency*. For an MIMD system to be efficient, memory bandwidth must be high enough to provide for simultaneous operation of all the processors. When memory modules are shared, the memory contention must be minimized. Also, the latency (which is the time elapsed between a processor's request for data from the memory and its receipt) must be minimized. This section examines memory organization techniques that reduce these problems to a minimum and tolerable level.

Memory Latency is reduced by increasing the memory bandwidth, which in turn is accomplished by one or both of the following mechanisms:

1. By building the memory system with multiple independent memory modules, thus providing for concurrent accesses of the modules. *Banked*, *interleaved*, and a combination of the two addressing architectures have been used in such systems. A recent trend is to use multiport memory modules in the design to achieve concurrent access.

2. By reducing the memory access and cycle times utilizing memory devices from the highest speed technology available. This is usually accompanied by high price. An alternative is to use cache memories in the design.

In order to understand the first method, consider an MIMD system with $N$ processors and a shared-memory unit. In the worst case all but one processor may be waiting to get access to the memory and not doing any useful computation, since only one processor can access the memory at a given instant of time. This bottlenecks the overall performance of the system.

A solution to this problem is to organize memory such that more than one simultaneous access to the memory is possible.

> **Example 6.2**
> Figure 6.3(a) shows an MIMD structure in which $N$ memory modules are connected to $N$ processors through a crossbar interconnection network. All the $N$ memory modules can be accessed simultaneously by $N$ different processors through the crossbar. In order to make the best possible use of such a design, all the instructions to be executed by one processor are kept in one memory module. Thus a given processor accesses a given memory block for as long a time as possible and the concurrency of memory accesses can be maintained over a longer duration of time. This mode of operation requires the banked memory architecture.
>
>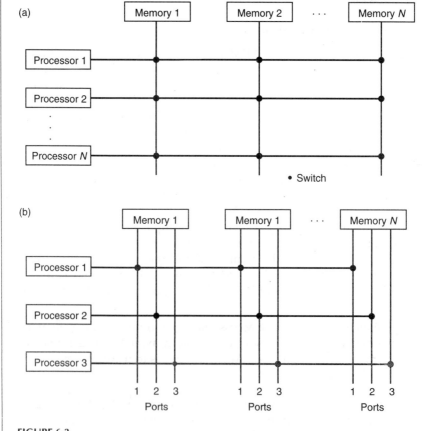
>
> **FIGURE 6.3**
> Processor–memory interconnection: (a) crossbar and (b) multiport memories

If an interleaved memory architecture is used, consecutive addresses lie in different memory modules. Thus the instructions corresponding to a task would be spread over several memory modules. If two tasks require the same code segment, it is possible to allow simultaneous access to the code segment, as long as one task starts slightly (at least one instruction cycle time) earlier than the other. Thus processors accessing the code march one behind the other spreading the memory access to different modules and minimizing contention.

Figure 6.3(b) shows the use of multiport memories. Here, each memory module is a three port memory device. All the three ports can be active simultaneously, reading and writing data to and from the memory block. The only restriction is that only one port can write data into a memory location. If two or more ports try to access the same location for writing, only the highest priority port will succeed. Thus, multiport memories have the contention resolution logic built into them and provide for concurrent access, at the expense of complex hardware. Large multiport memories are still expensive, because of their hardware complexity.

The architecture of Figure 6.2(c) depicts the second method for increasing memory bandwidth. Here, the cache memory is a high-speed buffer memory that is placed in close proximity to the processor (i.e., local memory). Anytime the processor wants to access something from the memory, it first checks its cache memory. If the required data is found there (i.e., a cache "hit"), it need not access the main (shared) memory which is usually 4 to 20 times slower. The success of this strategy depends on how well the application can be partitioned such that a processor accesses its private memory as long as possible (i.e., very high cache-hit ratio) and very rarely accesses the shared memory. This also requires that the interprocessor communication be minimized.

### 6.2.1 Cache Coherence

Consider the multiprocessor system of Figure 6.2(c) in which each processor has a local (private) memory. The local memory can be viewed to be a cache. As the computation proceeds, each processor updates its cache. But updates to a private cache are not visible to other processors. Thus, if one processor updates its cache entry corresponding to a shared data item, the other caches containing that data item will not be updated and hence the corresponding processors operate with stale data. This problem wherein the value of a data item is not consistent throughout the memory system is known as *cache incoherency*. Hardware and software schemes should then be applied to insure that all processors see the most recent value of the data item all the time. This is the process of making the caches coherent.

Two popular mechanisms for updating cache entries are *write-through* and *write-back*. In write-through, a processor updating the cache also simultaneously updates the corresponding entry in the main memory. In write-back, an updated cache-block is written back to the main memory just before that block is replaced in the cache.

The write-back mechanism clearly does not solve the problem of cache incoherency in a multiprocessor system. Write-through keeps the data coherent in a single-processor environment. But, consider a multiprocessor system in which processors 1 and 2 both load block A from the main memory into their caches. Suppose processor 1 makes some changes to this block in its cache and writes-through to the main memory. Processor 2 will still see stale data in its cache since it was not updated. Two possible solutions are

1. Update all caches that contain the shared data when the write-through takes place. Such a write-through will create an enormous overhead on the memory system.
2. Invalidate the corresponding entry in other processor caches when a write through occurs. This forces the other processors to copy the updated data into their caches when needed later.

Cache coherence is an active area of research interest. Several cache coherency schemes have evolved over the years. Some popular schemes are outlined below:

1. The least complex scheme for achieving cache coherency is not to use private caches. In Figure 6.4, the cache is associated with the shared-memory system rather than with each processor. Any memory write by a processor will update the common cache (if it is present in the cache) and will be seen by other processors. This is a simple solution but has the major disadvantage of high cache contention since all the processors require access to the common cache.

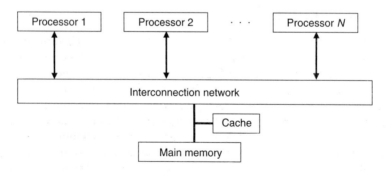

**FIGURE 6.4**
System with only one cache associated with the main (shared) memory

2. Another simple solution is to stay with the private cache architecture of Figure 6.4, but to only cache non-shared data items. Shared data items are tagged as non-cached and stored only in the common memory. The advantage of this method is that each processor now has its own private cache for non-shared data, thus providing a higher bandwidth. One disadvantage of this scheme is that the programmer and compiler has to tag data items as cached or non-cached. It would be preferred that cache coherency schemes were transparent to the user. Further, access to shared items could result in high contention.
3. *Cache flushing* is a modification of the previous scheme in which the shared data is allowed to be cached only when it is known that only one processor will be accessing the data. After the shared data in the cache is accessed and the processor is through using it, it issues a flush-cache instruction which causes all the modified data in the cache to be written back to the main memory and the corresponding cache locations to be invalidated. This scheme has the advantage of allowing shared areas to be cached; but it has the disadvantage of the extra time consumption for the flush-cache instruction. It also requires program code modification to flush the cache.
4. The above coherency schemes eliminate private caches, limit what may be cached or require programmer's intervention. A caching scheme which avoids these problems is *bus watching* or *bus snooping*. Bus snooping schemes incorporate hardware that monitors the shared bus for data LOAD and STORE into each processor's cache controller as shown in Figure 6.5. The snoopy cache controller controls the status of data contained within its cache based on the LOAD and STORE seen on the bus.

If the caches in this architecture are write-through, then every STORE to cache is written through simultaneously to the main memory. In this case, the snoopy controller sees all STOREs and take actions based on that. Typically,

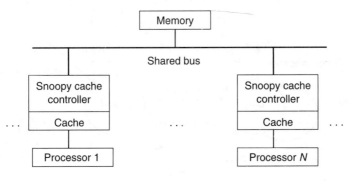

**FIGURE 6.5**
Snoopy bus controller on a shared bus

if a STORE is made to a locally cached block and the block is also cached in one or more remote caches, the snoopy controllers in remote caches will either update or invalidate the blocks in their caches. The choice of updating or invalidating remote caches will have its effect on the performance. The primary difference is the time involved to update cache entries versus merely changing the status of a remotely cached block. Second, as the number of processors increases the shared bus may become saturated. Note that for every STORE main memory must be accessed and for every STORE hit additional bus overhead is generated. LOADs are performed with no additional overhead.

For instance, the following explanation of the Illinois cache coherence protocol (Papamarcos and Patel, 1984) is based on the explanation in Archibald and Baer (1986). Each cache holds a cache state per block. The cache state is one of:

1. Invalid: the data for this block is not cached.
2. Valid-exclusive: the data for this block is valid, clean (identical to the data held in main memory), and is the only cached copy of the block in the system.
3. Shared: the data for this block is valid, clean, and there are possibly other cached copies of this block in the system.
4. Dirty: the data for this block is valid, modified relative to the main memory, and is the only cached copy of the block in the system.

    Initially, all blocks are invalid in all caches. Cache states change according to bus transactions as shown in Figure 6.6. Cache state transitions by the requesting processor are shown as solid lines. Cache state transitions of snooping processors are shown as dotted lines. For instance, on a read, if no snooping processor has a copy of the block, the requesting processor will transition to state valid-exclusive. If some snooping processor does have a copy of the block, then the

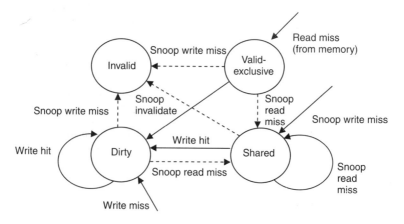

**FIGURE 6.6**
Illinois cache state transition diagram (Archibald and Baer, 1986)

requesting processor will transition to state shared. In addition, all snooping processors with a copy of the block will observe the read and transition to state shared. If some snooping processor has the block in state dirty, it will write the data to main memory at the same time.

5. A solution more appropriate for bus organized multiprocessor systems has been proposed by Goodman (1983). In this scheme, an invalidate request is broadcast only when a block is written in cache for the first time. The updated block is simultaneously written-through to the main memory. Only if a block in cache is written to more than once it is necessary to write it back before replacing it. Thus the first STORE causes a write-through to the main memory and it also invalidates the remotely cached copies of that data. Subsequent STOREs do not get written to the main memory, and since all other cache copies are marked invalid, copies to other caches are not required. When any processor executes a LOAD for this data, the cache-controller locates the unit (main memory or cache) that has a valid copy of the data. If the data is in a block marked dirty, then the cache supplies the data and writes the block to the memory. This technique is called *write-once*.

In a write-back cache scheme, since a modified cache block is loaded to the main memory only when that block is to be replaced, it conserves the bandwidth on the shared bus and thus is generally faster than write-through. But this added throughput is at the cost of a more complex bus-watching mechanism. In addition to the cache controller watching a bus, it must also maintain ownership information for each cached block, allowing only one copy of the cached block at a time to be writable. This type of protocol is called an *ownership protocol*.

An ownership protocol works in general, as follows. Each block of data has one owner. If main memory or a cache owns a block, all other copies of that block are Read-Only (RO). When a processor needs to write to an RO block, a broadcast to the main memory and all caches is made in an attempt to find any modified copies. If a modified copy exists in another cache, it is written to main memory, copied to the cache requesting read–write privileges, and then the privileges are granted to the requesting cache.

This section has addressed the most primitive cache coherency schemes. Cache coherence has been an active area of research and has resulted in several other schemes. (Refer to Eisner et al., 2000; Kim et al., 2002; Marathe et al., 2004 for further details).

## 6.3 Interconnection Networks

The interconnection network is an important component of an MIMD system. The ease and speed of processor-to-processor and processor-to-memory communication is dependent on the interconnection network design. A system can use either a static or a dynamic network, the choice depending on the dataflow and program characteristics of the application. The design, structure, advantages, and disadvantages of a number of interconnection networks were described in Chapter 5. This section extends that description to MIMD systems.

In a shared-memory MIMD system, the data exchange between the processors is through the shared memory. Hence an efficient memory to processor interconnection network is a must. This network interconnects the "nodes" in the system, where a node is either a processor or a memory block. A processor-to-processor interconnection network is also present in these systems. This network (more commonly called a synchronization network), provides for one processor to interrupt the other to inform that the shared data is available in the memory.

In a message-passing MIMD system, the interconnection network provides for the efficient transmission of messages between the nodes. Here, a "node" is typically a complete computer system consisting of a processor, memory, and I/O devices.

The most common interconnection structures used in MIMD systems are

1. Bus
2. Loop or ring
3. Mesh
4. Hypercube
5. Crossbar
6. Multistage switching networks

The following table lists representative MIMD systems and their interconnection networks.

| Interconnection network | Example system |
| --- | --- |
| Bus | Carnegie Mellon University Cm* |
| | Sequent Symmetry and Balance |
| Crossbar | Alliant FX/8 |
| Multistage | Carnegie Mellon University C.mmp |
| | BBN Butterfly |
| | IBM GF-11 and RP3 |
| Mesh (2D) | Intel Paragon XP/S |
| Hypercube | Intel iPSC |
| | nCUBE systems |
| Three-Dimensional Torus | Cray Research Inc. T3D |
| Tree | Columbia University DADO and NON-VON |
| | Thinking Machines CM-5 |

Details of Loop, Mesh, Hypercube, and Crossbar networks were provided in Chapter 5 as applied to SIMD systems. These networks are used in MIMD system design also, except that the communication occurs in an asynchronous manner, rather than the synchronous communication mode of SIMD systems. The rest of this section highlights the characteristics of these networks as applied to MIMD systems and covers bus and multistage switching networks in detail.

### 6.3.1 Bus Network

Bus networks are simple to build and provide the least cost among the three types of dynamic networks discussed in Chapter 5. They also offer the lowest performance. The bandwidth of the bus is defined as the product of its clock frequency and the width of the data path. The bus bandwidth must be large enough to accommodate the communication needs of all the nodes connected to it. Since the bandwidth available on the network for each node decreases as the number of nodes in the network increases, bus networks are suitable for interconnecting a small number of nodes.

The bus bandwidth can be increased by increasing the clock frequency. But technological advances that make higher bus clock rates possible also provide faster processors. Hence, the ratio of processor speed to bus bandwidth is likely to remain the same, thus limiting the number of processors that can be connected to a single bus structure.

The length of the bus also affects the bus bandwidth since the physical parameters such as capacitance, inductance, and signal degradation are proportional to the length of wires. Also, the capacitive and inductive effects grow with the bus frequency, thus limiting the bandwidth.

Figure 6.7 shows a shared memory MIMD system. The global memory is connected to a bus to which several nodes are connected. Each node consists of a processor, its local memory, cache, and I/O devices. In the absence of cache and local memories, all nodes try to access the shared memory through the single bus. For such a structure to provide maximum performance, both the shared bus and the shared memory should have high enough bandwidths. These bottlenecks can be reduced if the application is partitioned such that a majority of memory references by a processor are to its local memory and cache blocks, thus reducing the traffic on the common (shared) bus and the shared memory. Of course, the presence of multiple caches in the system brings in the problem of cache coherency.

If we use a multiport memory system for the shared memory, then a multiple-bus interconnection structure (see Figure 6.3(b)) can be used, with each port of the memory connected to a bus. This structure reduces the number of processors on each bus.

Another alternative is a *Bus window* scheme shown in Figure 6.8(a). Here, a set of processors is connected to a bus with a switch (i.e., a bus window) and all such buses are connected to form the overall system. The message transmission characteristics are identical to those of global bus, except that

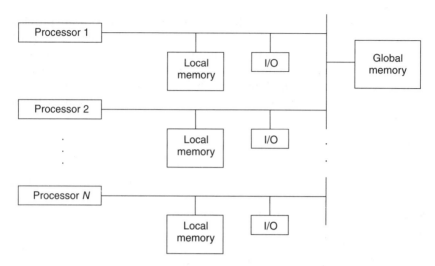

**FIGURE 6.7**
Shared-memories, shared-bus MIMD architecture

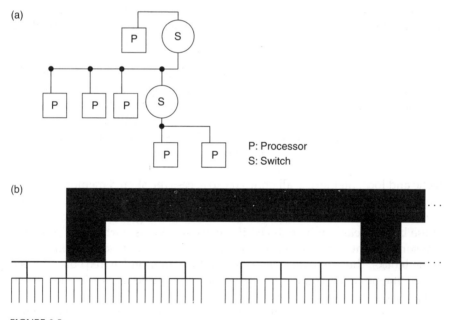

**FIGURE 6.8**
Interconnection structures: (a) bus window and (b) fat tree

multiple bus segments are available. Messages can be retransmitted over the paths on which they were received or on other paths.

Figure 6.8(b) shows a *Fat Tree* network that is gaining popularity. Here, communication links are fatter (i.e., have higher bandwidth) when they interconnect more nodes. Note that in practice, applications are partitioned such that the processes in the same cluster communicate with each other more often than with those in other clusters. As such, the links near the root of the tree must be thinner compared to the ones near the leaves. Thinking Machine Incorporated CM-5 uses the fat tree interconnection network.

Several standard bus configurations (Multibus, VME Bus, etc.) have evolved over the years. They offer support (in terms of data, address, and control signals) for multiprocessor system design.

### 6.3.2 Loop or Ring

The ring network is suitable for message-passing MIMD systems. The nodes are interconnected by a ring as in Figure 5.11(b), with a point-to-point interconnection between the nodes. The ring could be either unidirectional or bidirectional. In order to transmit a message, the sender places the message on the ring. Each node in turn examines the message header and buffers the message if it is the designated destination. The message eventually reaches the sender, which removes it from the ring.

One of the popular protocols used in rings is the *token ring* (IEEE 802.5) standard. A token (which is a unique bit pattern) circulates over the ring. When a node wants to transmit a message, it accepts the token (i.e., prevents it from moving to the next node) and places its message on the ring. Once the message is accepted by the receiver and reaches the sender, the sender removes the message and places the token on the ring. Thus, a node can be a transmitted only when it has the token.

Since the interconnections in the ring are point-to-point, the physical parameters can be controlled more readily, unlike bus interconnections, especially when very high bandwidths are needed.

One disadvantage of token ring is that each node adds a 1-bit delay to the message transmission. Thus, the delay increases as the number of nodes in the system increases. If the network is viewed as a pipeline with a long delay, the bandwidth of the network can be effectively utilized. To accommodate this mode of operation, the nodes usually overlap their computations with the message transmission.

One other way to increase the transmission rate is to provide for the transmission of a new message as soon as the current message is received by the destination node, rather than waiting until the message reaches the sender, where it is removed.

### 6.3.3 Mesh Network

The Mesh networks are ideal for applications with very high near-neighbor interactions. If the application requires a large number of global interactions, the efficiency of the computation goes down, since the global communications require multiple hops through the network. One way to improve the performance is to augment the mesh network with another global network. MasPar architectures and Intel iPSC architectures utilize such global interconnects.

### 6.3.4 Hypercube Network

One advantage of the hypercube networks is that routing is straightforward and the network provides multiple paths for message transmission from each node. Also, the network can be partitioned into hypercubes of lower dimensions and hence multiple applications utilizing smaller networks can be simultaneously implemented. For instance, a four-dimensional hypercube with 16 processing nodes can be used as two three-dimensional hypercubes, four two-dimensional hypercubes, etc.

One disadvantage of the hypercube is its scalability, since the number of nodes has to be increased in powers of two. That is, to increase the number of nodes from 32 to 33, the network needs to be expanded from a five-dimensional to a six-dimensional network consisting of 64 nodes. In fact, the Intel Touchstone project has switched over to mesh networks from hypercubes, due to this scalability issue.

### 6.3.5 Crossbar Network

The crossbar network (Figure 5.15) offers multiple simultaneous communications with the least amount of contention, but at a very high hardware complexity. The number of memory blocks in the system is at least equal to the number of processors. Each processor to memory path has just one crosspoint delay.

The hardware complexity and the cost of the crossbar is proportional to the number of crosspoints. Since there are $N^2$ crosspoints in an $(N \times N)$ crossbar, the crossbar becomes expensive for large values of $N$.

### 6.3.6 Multistage Networks

Multistage switching networks offer a cost/performance compromise between the two extremes of bus and crossbar networks. A large number of multistage networks have been proposed over the past few years. Some examples are Omega, Baseline, Banyan, and Benes networks. These networks differ in their topology, operating mode, control strategy, and the type of switches used. They are capable of connecting any source (input) node to any destination (output) node. But, they differ in the number of different $N$-to-$N$ interconnection patterns they can achieve. Here, $N$ is the number of nodes in the system.

These networks are typically composed of 2-input, 2-output switches (see Figure 6.9) arranged in $\log_2 N$ stages. Thus the cost of the network is of

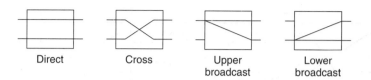

**FIGURE 6.9**
2 × 2 Crossbar

the order of $N \log_2 N$ compared to $N^2$ of the crossbar. Communication paths between nodes in these networks are of equal length (i.e., $\log_2 N$). The latency is thus $\log_2 N$ times that of a crossbar, in general. In practice, large crossbars have longer cycle times compared to those of the small switches used in multistage networks.

The majority of multistage networks are based on the perfect shuffle interconnection scheme described in Chapter 5. Two popular networks in this class are the Omega and Benes networks. *Omega Network* (Almasi and Gottlieb, 1989).

The Omega network is the simplest of the multistage networks. An $N$-input, $N$-output Omega interconnection topology is shown in Figure 6.9. It consists of $\log_2 N$ stages. The perfect-shuffle interconnection is used between the stages. Each stage contains $N/2$ 2-input, 2-output switches. These switches can perform the four functions shown in Figure 6.9.

The network employs the packet switching mode of communication, where each packet is composed of data and the destination address. The address is read at each switch and the packet is forwarded to the next switch until it arrives at the destination node. The routing algorithm follows a simple scheme. Starting from the source, each switch examines the leading bit of the destination address and removes that bit. If the bit is 1, then the message exits the switch from the lower port; otherwise, from the upper port.

Figure 6.10 shows an example in which the message is sent from node (001) to node (110). Since the first bit of the destination address is 1, the switch in the first stage routes the packet to its lower port. Similarly, the switch in the second stage also routes it to its lower port since the second bit of the address is also 1. The switch in the third stage routes it to its upper port since the third bit of the address is 0. The path is shown as solid line in the figure.

The above routing algorithm implies a distributed control strategy, since the control function is distributed among the switches. An alternative control strategy would be centralized control, in which the controller sends control signals to each switch based on the routing required. It may operate either on a stage by stage basis, starting from the first to the last or it may reset the whole network each time, depending on the communication requirements.

Figure 6.10 also shows (in dotted lines) the path from node (101) to node (100). Note that the link between stages 1 and 2 is common to both the paths

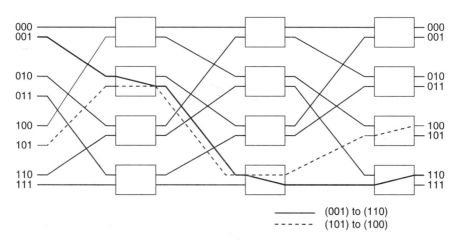

**FIGURE 6.10**
Omega network

shown in the figure. Thus, if nodes (001) and (101) send their packets simultaneously to the corresponding destination nodes, one of the packets is "blocked" until the common link is free. Hence, this network is called a "blocking" network.

Note that there is only one unique path in this network from each input node to an output node. This is a disadvantage since the message transmission gets blocked, even if one link in the path is part of the path for another transmission in progress. One way to reduce the delays due to blocking is to provide buffers in switching elements so that the packets can be retained locally until the blocked links are free. The switches can also be designed to "combine" messages bound for the same destination. Recall that a crossbar is a nonblocking network, but it is much more expensive than an omega network.

### 6.3.6.1 Benes network (Almasi and Gottlieb, 1989)

Unlike the Omega network which has a single unique path between a source and a destination node, Benes network provides multiple paths between the source and the destination nodes. Figure 6.11 shows the 8 × 8 Benes network. Note that the number of stages has increased to 5 and each stage contains four 2-input, 2-output switches. The paths available for packet transmission from node 010 to node 110 are shown in Figure 6.11(a). The two blocking connections in the Omega network of Figure 6.11 can be simultaneously made in this network as shown in Figure 6.11(b).

The routing algorithm for Benes network is not as simple as that for the Omega network. Each path needs to be determined independently, before the transmission begins. When multiple simultaneous transmissions are required, the network can achieve most of the permutations if all the

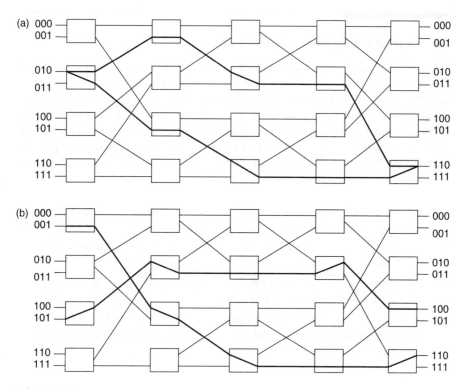

**FIGURE 6.11**
8 × 8 Benes network: (a) alternate paths from (010) to (110) and (b) simultaneous paths (001) to (110) and (101) to (100)

connections required are known before the routing computation begins. If all the desired connections are not known at the beginning, the network rearranges some routings made earlier to accommodate new requests. Hence, it is called a "rearrangeble" network.

Benes network offers the advantages of reduced blocking and increased bandwidth over Omega network. But, it requires more complex hardware and increases the network latency because of the increased number of stages. It is suitable for circuit-switched environments since the routing computations take a considerable amount of time. A Benes network is used by the IBM GF11 to interconnect 576 processors. Each switch in this network is a 24 × 24 crossbar.

The progress in hardware technology has resulted in the availability of fast processors and memory devices, two components of an MIMD system. Standard bus systems such as Multibus and VME bus allow the implementation of bus-based MIMD systems. Although hardware modules to implement other interconnections (loop, crossbar, etc.) structures are now appearing off-the-shelf, the design and implementation of an appropriate interconnection structure is still the crucial and expensive part of the MIMD design.

## 6.4 Operating System Considerations

In an MIMD system, several processors are active simultaneously. Each active processor is associated with one or more memory blocks depending on data and program characteristics. The I/O devices are either dedicated to individual processors or are shared by all the processors. The overall working of the system is coordinated by a multiprocessor operating system, whose functions are very similar to those of an SISD operating system, except for a much higher complexity. The major functions of the multiprocessor operating system are

1. Keeping track of the status of all the resources (processors, memories, switches, and I/O devices) at any instant of time.
2. Assigning tasks to processors in a justifiable manner (i.e., to maximize processor utilization).
3. Spawning or creating new processes such that they can be executed in parallel or independent of each other.
4. When all the spawned processes are completed, collecting their individual results and passing them to other processes as required.

---

**Example 6.3**

Consider for example, the element by element addition of two vectors **A** and **B** to create the vector **C**. That is,

$$c_i = a_i + b_i \quad \text{for } i = 1 \text{ to } n \tag{6.1}$$

It is clear that the computation consists of $n$ additions that can be executed independent of each other. Thus an MIMD with $n$ processors can do this computation in one addition time. If the number of processors $m$ is less than $n$, the first $m$ tasks are allocated to the available processors and the remaining tasks are held in a queue. When a processor completes the execution of the task allocated to it, a new task (from the queue) is allocated to it. This mode of operation continues until all the tasks are completed, as represented by the following algorithm:

1. /* Spawn $n-1$ processes each with a distinct process identification number $k$. Each spawned process starts at 'label' */

    ```
    for k = 1 to n-1
       FORK label(k);
    ```

2. /* the process that executed FORK is assigned $k = n$. This is the only process that reaches here, the other spawned processes jump directly to 'label' */

   ```
   k = n;
   ```

3. /* add $k^{th}$ element of each vector; $n$ different processes perform this operation, not necessarily in parallel */

   ```
   label: c[k] = a[k] + b[k];
   ```

4. /* terminate the $n$ processes created by FORK; only 1 process continues after this point */

   ```
   JOIN n;
   ```

The new aspects of this algorithm are the FORK and JOIN constructs. They are two of the typical commands to any multiprocessing operating system used to create and synchronize tasks (processes). The FORK command requests the operating system to create a new process with a distinct process identification number ($k$ in this example). The program segment corresponding to the new process starts at the statement marked "label."

Note that initially the entire algorithm above constitutes one process. It is first allocated to one of the free processors in the system. This processor through the execution of the FORK-loop (step 1) requests the operating system to create $(n-1)$ tasks, after which it continues with step 2. Thus, after the execution of step 2, there are $n$ processes in all waiting to be executed. The process with $k = n$ continues on the processor that is already active (i.e., the processor that spawned the other processes). The remaining $n-1$ processes created by the operating system, enter a process queue. The processes waiting in the queue are allocated processors as processors become available.

In this example, the $k$th process adds $k$th elements of A and B creating the $k$th element of C. The program segment corresponding to each process ends with the JOIN statement. The JOIN command can be viewed as the inverse of the FORK command. It has a counter associated with it that starts off at 0. A processor executing the JOIN increments the counter by 1 and compares it to $n$. If the value of the counter is not equal to $n$, the processor cannot execute any further and hence it terminates the process and returns to the available pool of processors for subsequent allocation of tasks. If the value of the counter is $n$, the process, unlike the others that were terminated, continues execution beyond the JOIN command. Thus, JOIN ensures that the $n$ processes spawned earlier have been completed before proceeding further in the program.

Several aspects of the above algorithm are worth noting:

1. This algorithm does not use the number of processors $m$ as a parameter. As such it works on MIMD systems with any number of processors. The overall execution time depends on the number of processors available.
2. Operating system functions of creating and queuing tasks, allocating them to processors, etc. are done by another processor (or a process) which is not visible in the above discussion.
3. The procedure for creating processes requires $O(N)$ time. This time can be reduced to $O(\log_2 N)$ by a more complex algorithm that makes each new process perform fork, thus executing more than one fork simultaneously.
4. Step 1 of the algorithm was executed only $n - 1$ times, thus the $n$th process was not specifically created by a FORK, but was given to the processor that was active already. Alternatively, the FORK-loop could have been executed $n$ times. Then the processor executing the loop must have been deallocated and brought to the pool of available processors. A new process could have then been allocated to it. Thus, the above procedure eliminates the overhead of deallocation and allocation of one process to a processor. Typically, creation and allocation of tasks results in a considerable overhead requiring execution of about 50 to 500 instructions.
5. Each process in this example performs an addition and hence is equivalent to three instructions (Load A, Add B, Store C). Thus, the process creation and allocation overhead mentioned above is of the order of 10 to 100 times the useful work performed by the process. As such, the above computation would be more efficient on an SISD rather than an MIMD. But the example serves the purpose of illustrating the operation of an MIMD system.

### Example 6.4

As another example, the implementation of matrix multiplication algorithm on MIMD is shown below.

```
/* Input: 2 N x N   matrices A and B */
/* Output: Product matrix C */
      for k = 1 to n-1    /*spawn n-1 processes    */
          FORK label(k);
      endfor;
          k = n;       /*the nth process     */
/* n different processes with distinct
      process id numbers reach this point */
```

```
label: for i = 1 to n
       C[i,k] = 0;
       for j = 1 to n
       C[i,k] = C[i,k] + A[i,j] * B[j,k];
       endfor;
       endfor;
       JOIN n; /* terminate the N processes and gather
               results */
```

Figure 6.12 shows how this algorithm is executed on an MIMD. Here, matrices **A** and **B** are assumed to be of the order $5 \times 5$ and the MIMD has four processors. The task is first allocated to Processor 2 which executes the FORK loop to spawn other processes while other processors are idle. The first process (corresponding to $k = 1$) is allocated to Processor 3, the second ($k = 2$) to Processor 1, and the third ($k = 3$) to Processor 4. After spawning the fourth process ($k = 4$), Processor 2 continues to execute the process $k = 5$, while process $k = 4$ is placed in the queue. Processor 3 completes the task ($k = 1$) allocated to it first. Since the other processes have not reached the JOIN yet, Processor 3 cannot continue and hence joins the available processor pool. Process $k = 4$ waiting in the queue is then allocated to it. Meanwhile Processors 1, 2, and 4 complete their tasks and execute the JOIN. Processor 3 is the last to execute JOIN (i.e., when the JOIN counter reaches 5) and thus continues execution beyond the JOIN instruction.

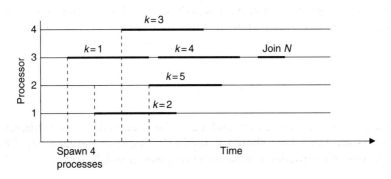

**FIGURE 6.12**
Execution-time characteristics

Again, it is assumed that the operating system functions are performed by a fifth processor not visible in the above description. Further details on fork and join constructs are provided later in this chapter.

In the above example, a task consists of the code segment starting at "label" and ending at JOIN. For this task to execute independently of other tasks, the data required by it (matrix **A** and the *k*th column of matrix **B**) should be available for its exclusive use. That means, the code segment should be duplicated for each task and corresponding number of copies of matrix **A** should be made. This duplication increases the memory requirements tremendously and hence is impractical for large values of $n$. One solution would be to organize the memory in an interleaved fashion and let all the tasks share the single copy of the code segment and data segment corresponding to the **A** matrix, although in a delayed manner (i.e., each task is started slightly later than the previous one, so that the memory references are distributed to several physical memory modules). Of course, this does not guarantee a contention-free operation always, although it minimizes the memory contention.

### 6.4.1 Synchronization Mechanisms

There was no explicit interprocess communication in the above example. In practice, the various processes in the multiprocessor system need to communicate with each other. Since the processes executing on various processors are independent of each other and the relative speeds of execution of these processes cannot be easily estimated, a well-defined synchronization between processes is needed for the communication (and hence the results of computation) to be correct. That is, processes in an MIMD operate in a *cooperative* manner and a *sequence control* mechanism is needed to ensure the ordering of operations. Also, processes *compete* with each other to gain *access* to shared data items (and other resources). An *access control* mechanism is needed to maintain orderly access. This section describes the most primitive synchronization techniques used for access and sequence control.

---

**Example 6.5**

Consider two processes P1 and P2 to be executed on two different processors. Let S be a shared variable (location) in the memory. The sequence of instructions in the two processes is as follows:

```
P1:   1.  MOV Reg1,[S]  /* The first operand is the
                           destination */
      2.  INC Reg1      /* Increment Reg1 */
      3.  MOV [S],Reg1

P2:   1'. MOV Reg2,[S]
      2'. ADD Reg2,#2   /* Add 2 to Reg2 */
      3'. MOV [S],Reg2
```

The only thing we can be sure of as far as the order of execution of these instructions is concerned is that instruction 2(2′) is executed after instruction 1(1′) and instruction 3(3′) is executed after instructions 1(1′) and 2(2′). Thus the following three cases for the actual order of execution are possible:

1. 1 2 3 1′2′3′ or 1′2′3′1 2 3: location S finally has the value 3 in it. (Assuming $S = 0$, initially)
2. 1 1′2 2′3 3′: location S finally has value 2 in it
3. 1 1′2′2 3′3: location S finally has value 1 in it

The desired answer is attained only in the first case and that is possible only when process P1 is executed in its entirety before process P2 or vice versa. That is, P1 and P2 need to execute in a *mutually exclusive* manner. The segments of code which are to be executed in a mutually exclusive manner are called *critical sections*. Thus in P1 and P2 of this example, the critical section is the complete code corresponding to each process.

**Example 6.6**
As another example consider the code shown in Figure 6.13 to compute the sum of all the elements of an $n$-element array A. Here, each of the $n$ processes spawned by the fork loop, adds an element of A to the shared variable SUM. Obviously, only one process should be updating SUM at a given time. Imagine process 1 reads the value of SUM and has not yet stored the updated value. Meanwhile, if process 2 reads SUM and updates are done by process 1 followed by process 2, the resulting SUM would be erroneous. To obtain the correct value of SUM, we should make sure that once process 1 reads SUM, no other process can access SUM until process 1 writes the updated value. This is the mutual exclusion of processes.

The mutual exclusion is accomplished by LOCK and UNLOCK constructs. This pair of constructs has a flag associated with it. When a process executes LOCK, it checks the value of the flag. If the flag is ON, it implies that some other process has accessed SUM and hence the process waits until the flag is OFF. If the flag is not ON the process sets it ON and gains access to SUM, updates it and then executes the UNLOCK which clears the flag. Thus the LOCK/UNLOCK brings about the synchronization of the processes.

```
SUM = 0;

for k = 1 to n-1    /*spawn n-1 processes    */

    FORK label(k);

endfor;

k = n;              /*the nth process    */

/*n different processes with distinct process id

        numbers reach this point        */

    label: LOCK (flag)

        SUM = SUM + A(k);

        UNLOCK (flag);

        JOIN n;  /* terminate the N processes and gather

            results    */
```

**FIGURE 6.13**
Synchronization using LOCK/UNLOCK

Note that during the LOCK operation, the functions of fetching the flag, checking its value, and updating it must all be done in an *indivisible* manner. That is, no other process should have access to the flag until these operations are complete. Such indivisible operation is brought about by a hardware primitive known as TEST_AND_SET.

### 6.4.1.1    TEST_AND_SET

The use of TEST_AND_SET primitive is shown in Figure 6.14. Here, $K$ is a shared-memory variable which can have a value of either 0 or 1. If it is 0, the TEST_AND_SET returns a 0 and sets $K$ to 1. The process enters its critical section. If $K$ is 1, TEST_AND_SET returns 1 thus locking the process from entering the critical section. When the process is through executing its critical section, it resets $K$ to 0 thus allowing a waiting process access into its critical section.

In Figure 6.13, the critical section is not part of the While loop. The range of the While loop is a single statement (terminated by ";"). The While loop makes the process wait until $K$ goes to 0, to enter the critical section. There are two modes of implementing the wait: busy waiting (or spin lock) and task

# Multiprocessor Systems

```
P1:  while not(TEST_AND_SET(K));

        ┌──────────────────────┐
        │ critical section of P1; │
        └──────────────────────┘

     K = 0;

P2:  whilenot(TEST_AND_SET(K));

        ┌──────────────────────┐
        │ critical section of P2; │
        └──────────────────────┘

     K = 0;


     /* The body of TEST_AND_SET procedure */

     TEST_AND_SET(K)
     {
         temp = K;
         K = 1;
         return(temp);
     }
```

**FIGURE 6.14**
Use of TEST_AND_SET

switching. In the first mode the process stays active and repeatedly checks the value of $K$ until it is 0. Thus if several processes are busy-waiting, they keep the corresponding processors busy but no useful work gets done. In the second mode, the blocked process is enqueued and the processor is switched to a new task. Although this mode allows better utilization of processors, the task-switching overhead is usually very high, unless special hardware support is provided.

Several processes could be performing the TEST_AND_SET operation simultaneously, thus competing to access the shared resource $K$. Hence, the process of examining $K$ and setting it must be indivisible in the sense that $K$ cannot be accessed by any other process until the TEST_AND_SET is completed once. TEST_AND_SET is usually an instruction in the instruction set of the processor and is the minimal hardware support needed to build other high-level synchronization primitives.

The TEST_AND_SET effectively *serializes* the two processes so that they execute in a mutually exclusive manner. Once a process is in the critical section, other processes are *blocked* from entering it by the TEST_AND_SET. Thus, the mutual exclusion is brought about by *serializing* the execution of processes and hence, affecting the parallelism. Also, the blocked processes incur overhead due to busy-waiting or task switching.

```
                    P1: P(S);

                         ┌─────────────────┐
                         │ critical section; │
                         └─────────────────┘

                    V(S);

                    P2: P(S);

                         ┌─────────────────┐
                         │ critical section; │
                         └─────────────────┘

                    V(S);
```

**FIGURE 6.15**
Synchronization using P and V

### 6.4.1.2  Semaphores

Dijkstra (1968) introduced the concept of semaphores and defined two high-level synchronization primitives P and V based on the *semaphore* variable S. They are defined below:

*P(S) or WAIT(S)*

If $S = 0$, the process invoking P is delayed until $S > 0$.

If $S > 0$, $S = S - 1$ and the process invoking P enters the critical section.

*V(S) or SIGNAL(S)*

$$S = S + 1$$

S is initialized to 1. The testing and decrementing of S in P(S) is indivisible. So is the incrementing of S in V(S). Figure 6.15 shows the use of P and V to synchronize processes P1 and P2 of Figure 6.14.

S is a binary variable in the implementation of P and V above. If S is an integer variable it is called a counting semaphore. If it is initialized to M, then M processes can be in the critical section at any time.

### 6.4.1.3  Barrier synchronization

Consider the following subprogram running on an MIMD in a parallel manner:

```
for i = 1 to n
{
    step A[i];
    step B[i];
    step C[i];
}
```

Here, the "step" refers to a computation that generates the value for the corresponding variable. This subprogram is divided say into $n$ processes, with each independent process taking care of what would be one iteration of the loop in a sequential environment. Due to the nature of MIMD performance, nothing can be predicted about the order of these steps on various processors. The only thing we are sure of is that for a given value of $i$ (i.e., for a given process), the steps will be in the order A followed by B followed by C. Thus, A[1],A[2],B[1],C[1],B[2],C[2] and A[2],B[2],C[2],A[1],B[1],C[1] are both valid sequences.

Suppose the algorithm requires that the results from steps A and B of all iterations are needed in step C of the iterations. This is possible if and only if all the processes execute their steps A and B before any process starts its step C. This is assured by the inclusion of a BARRIER command after step B in the algorithm, as shown below:

```
for i = 1 to n
{
    step A[i];
    step B[i];
    BARRIER;
    step C[i];
}
```

In this method of synchronization, the processes must wait for the slowest process to reach the barrier before continuing. Thus, if the slowest process executes much longer than the others, most of the processors are tied up busy-waiting. Also, if the number of processes busy-waiting is greater than the number of processors available, a *deadlock* occurs. A deadlock occurs when there exists a cycle of precedence such that A is waiting for B, B is waiting for C, and so on, with the last element in the cycle waiting for A. None of the processes can continue to execute unless the deadlock is broken, by aborting one or more of them.

### 6.4.1.4   FETCH_AND_ADD

FETCH_AND_ADD is similar to TEST_AND_SET in its implementation, but is *nonblocking* synchronization primitive. That is, it allows the operation of several processes in the critical section in parallel yet nonconflicting manner. FETCH_AND_ADD is shown below:

```
FETCH_AND_ADD(S, T)
    { Temp = S;
            S = S + T;   }
Return Temp;
```

Two parameters are passed to FETCH_AND_ADD: $S$ the shared variable and $T$ an integer. If $S$ is initialized to 0 and two processes P1 and P2 make a call to FETCH_AND_ADD at roughly the same time, the one reaching the FETCH_AND_ADD first receives the original value of $S$ and the second one receives $(S + T)$. The two processes can execute further independently, although the updating of $S$ is effectively serialized. In general, the FETCH_AND_ADD gives each contending process a unique number and allows them to execute simultaneously, unlike the TEST_AND_SET which serializes the contending processes.

As another example, the instruction FETCH_AND_ADD(SUM, INCREMENT) provides for the addition of INCREMENT to SUM by several processes simultaneously and results in the correct value of the SUM without the use of LOCK/UNLOCK.

FETCH_AND_ADD is useful for cases in which the same variable is accessed by several contending processes. FETCH_AND_ADD on different variables are done sequentially if those variables reside in the same memory. The implementation cost of FETCH_AND_ADD is high. As such it is limited to environments in which updates become a bottleneck because of tens of processes are contending for access to the shared variable. If only one process requests access to the shared variable at a time, TEST_AND_SET is more economical.

The processes in a message-passing MIMD are automatically synchronized since a message cannot be received before it is sent. Message-processing protocols deal with the problems of missing or overwritten messages and sharing of resources.

This section has described the most primitive synchronizing mechanisms. (Refer to Silberschatz et al., 2003; Stallings, 2002; Tanenbaum, 2001 for further details.)

### 6.4.2 Heavy- and Light-Weight Processes

Consider the UNIX operating system in which a new process is created by forking an existing process. When a fork is executed, the running program is interrupted (and usually swapped to disk) and two copies of the program are swapped back into the memory. These two copies are identical in all aspects except that the programmer uses their process ID to determine which process code is executing at that time. A process consists of program code, data, page tables, and file descriptors. The address space of a process consists of three segments: code, data, and stack. Processes do not share addresses and hence are mutually independent. An attempt by one process to access any segment of the other process is trapped as a fatal error. Creation of tasks of this type results in a high overhead and hence these processes are called *heavy-weight* processes. This mode of operation is good when the tasks are heterogeneous and is usually called *multitasking*.

In parallel programming environments, the tasks tend to be homogeneous. They share data segments between them and the high overhead of task creation is not desirable. For these reasons, extensions to UNIX have created the concept of *light-weight* processes and this mode of operation is termed *microtasking*.

For example, the Sequent Computer Corporation uses DYNIX, an extension to UNIX on their Balance series of multiprocessors. A process in DYNIX consists of the code, data, and stack segments as in UNIX. In addition, a shared data segment is also created. Subsequent microtasks spawned by the task copy all but the shared data segment into their address spaces. Thus, only one copy of shared data segment exists among all the worker tasks.

This model of execution requires high-level language (HLL) primitives to declare data elements as shared or private and two types of synchronization mechanisms are needed, one for microtasking and the other for multitasking. Some of the DYNIX primitives are listed below:

m_fork: Execute a subprogram in parallel

m_set_procs: Set the number of processes to be created by a subsequent m_fork.

m_kill_procs: Kill child processes.

m_lock, m_unlock: lock and unlock variables of the type slock_t (microtasking).

m_sync: Wait for all microtasks to arrive at the barrier.

m_single: Make the child processes to spin until the parent process has executed the code following m_single call and calls the m_multi.

m_multi: Resume all worker microtasks in parallel.

s_init_lock, s_lock, s_unlock: Initialize, lock and unlock, lock variables of the type slock_t (Multitasking).

s_init_barrier: Initialize the barrier.

s_wait_barrier: Wait for all Multitasks to arrive.

The following code segment from (Kallstrom and Thakkar, 1988) illustrates the use of the above primitives. It creates one supervisor and $(n-1)$ workers, where each worker's task is the function work().

```
shared int x;
shared struct y_struct {
        int state;
        slock_t lp;        /*primitive lock*/
           } y
/* Supervisor */
main() {
        s_init_lock (&y.lp); /*initialize lock*/
        m_set_procs(n);      /*initialize n_1 workers*/
        m_fork(work);
        m_kill_procs();      /*done*/
         }
```

```
/* Workers */
work(){
    /* Microtask locking */
    m_lock();              /* Mutual exclusive access to x */
    x++;
    printf ("Worker");
    m_unlock();            /* End of critical section */
    /* A barrier */
    m_single();            /* Workers wait here */
    printf ("Supervisor")  /* Only Supervisor does this */
    m_multi();             /* Workers resume */
    /* Multitask lock */
    s_lock(&y.lp);         /* lock the struct y */  y.state = o;
    printf ("Structure");
    s_unlock(&y.lp):       /* Unlock */
    }
```

Refer to DYNIX manuals for further details.

### 6.4.3 Scheduling

Recall that a parallel program is a collection of tasks. These tasks may run serially or in parallel. An *optimal schedule* determines the allocation of tasks to processors of the MIMD system and the execution order of the tasks, so as to achieve the shortest execution time. The scheduling problem in general is NP-complete. But, several constrained models have evolved over the years and currently this problem is an active area of research in parallel computing.

Scheduling techniques can be classified into two groups: static and dynamic. In *static* scheduling, each task is allocated to a particular processor based on the analysis of the precedence constraints imposed by the tasks at hand. Each time the task is executed, it is allocated to that predetermined processor. Obviously, this method does not take into consideration the non-deterministic nature of tasks brought about by conditional branches and loops in the program. The target of the conditional branch and the upper bounds of the loops are not known until the program execution begins. Thus, static scheduling will not be optimal.

In *dynamic* scheduling, tasks are allocated to processors based on the execution characteristics. Usually some *load balancing* heuristic is employed in determining optimal allocation. Since the scheduler has only the knowledge of local information about the program at any instant of time, finding the global optimum is difficult. Another disadvantage is the increased overhead since the schedule has to be determined while the tasks are running. (Refer to Adam et al. (1974), Bashir et al. (1983), and El-Rewini and Lewis (1990) for further details.)

## 6.5 Programming

The basic idea behind parallel processing is that a job is divided up and parceled out to multiple processors to form a cooperative effort. This chapter has so far concentrated on how the hardware and operating system software incorporate facilities for parallelism. These facilities can be exploited in full only if there exist techniques for specifying to the hardware what all is to be executed in parallel, in a given job.

Parallelism refers to the simultaneous execution of tasks, task steps, programs, routines, subroutines, loops, or statements. Five levels of parallelism can be identified:

1. Parallelism at the level of independent tasks and programs, that is, totally different programs running on different processors at any given instant of time.
2. Parallelism at the level of task steps and related parts of a program, that is, different parts of the same program are made to execute on different processors.
3. Parallelism at the level of routines and subroutines, that is, a number of different subroutines (programs or functions) of a given program are made to execute on different processors in a parallel manner.
4. Parallel execution of loops, that is, given a loop construct in a program, the iterations of this loop may be executed in parallel on different processors.
5. Parallel execution of statements and instructions, that is, individual statements forming a program are parceled out to different processors to be executed in parallel.

Note that the higher the level, the finer is the granularity of the software process. As pointed out by Ghezzi (1982), a parallel program must have the ability to:

1. Define the set of subtasks to be executed in parallel. These subtasks could be at any of the above five levels.
2. Start and stop the execution of subtasks.
3. Coordinate and specify the interaction between subtasks while they are executing.

A sequential program (i.e., a program for an SISD) has four main constructs. For each of these constructs a corresponding parallel construct can be identified as shown below:

1. Sequential one line statements executed one after the other. The corresponding parallel construct is the *parbegin/parend* which brackets

a set of statements for concurrent execution as in Concurrent Pascal and Algol 68. The statements occurring between parbegin and parend will not be necessarily executed in the order in which they appear in the program; instead they will be distributed over a number of processors which will execute them independently. Eventually all the results will be collected and passed on to the statement following the parend construct. An example showing the function of parbegin/parend construct is provided later in this section.

2. In order to make the logical order of execution different from the physical order of the statements in a sequential program, goto (unconditional) and if_then_else (conditional) constructs are used. The analog of these constructs for parallel programs is the *fork/join* pair of statements discussed earlier. An example of a language which uses a variant of fork/join is PL/1.

3. The simple looping construct (for, do) in case of sequential programs is comparable to the *doall* construct for parallel programs which is used to designate a block of nearly identical statements to be executed in parallel. In the best case, if there are N iterations in a loop, all these N iterations will be allocated to N different processors and executed simultaneously, thus attaining an N-fold speedup.

**Example 6.7**
Consider again the problem of performing the element by element addition of two linear arrays, A[1...N] and B[1...N]. A for-loop solution for this problem is as follows:

```
for i = 1 to N do
  sum[i] := A[i] + B[i];
endfor
```

This solution is satisfactory for a purely sequential environment. Running this on an MIMD system requires that the N summing tasks are parceled out to as many processors as possible and performed in parallel. This intention is explicitly indicated in the following program segment:

```
forall i = 1 to N do
  sum[i] := A[i] + B[i];
```

The keyword *forall* implies that the N processes are created and distributed among the processors which execute them in parallel. Considering the best case, N processors get these N processes at a time when all of them are free. Obviously all of the processes will be executed together and the total time taken will be equal to the time taken for one iteration of the sequential for loop, yielding a speedup of N.

4. In sequential programs, subprograms (procedures or functions) can be called more than once by the main and other modules. Correspondingly, processes, tasks, procedures, etc. can be declared for parallel execution (as in Concurrent Pascal and Ada).

Partitioning an application into subtasks for parallel execution on an MIMD is not trivial. There are efforts to device compilers that recognize the parallelism in a sequential program and partition it for a particular MIMD architecture, although such partitioning is not yet completely automatic and requires the interaction of the application designer with the compiler. There are also efforts to design languages that allow explicit specification of parallel tasks in the application.

It is possible to extend existing sequential languages to accommodate MIMD programming. The additional constructs needed for programming shared-address-space MIMD systems are: the primitives for mutual exclusion and synchronization (described in the previous section), the primitives for creating processes and the primitives to allocate shared variables. For programming message-passing MIMD systems, the constructs needed are: the message SEND and RECEIVE. The rest of this section provides further details on these parallel program constructs.

### 6.5.1 Fork and Join

The code segment in Figure 6.16 generalizes the fork/join construct. Here, A through F, each represent a set of sequential statements in the code. The whole code segment in the figure, is the only process to begin with. The first fork instruction, Fork c spawns a new process, which starts execution from the statement labeled c, while the original process continues into B. Similarly, Fork e and Fork f spawn two more new processes.

Each time a process encounters the Join $m, g$ instruction, it decrements $m$ by 1 and jumps to label $g$ if $m$ is zero after the decrement operation; the process cannot continue if $m$ is nonzero. The two processes that do this are C and D. After both these processes execute the Join $m, g$ the value of $m$ reaches zero (since it is initialized to 2), thus joining the two processes to form a single process, G. Note that the order in which the join statement is executed does not matter and in general cannot be predicted. The first process executing the join is terminated (since all the processes controlled by $m$ have not completed yet) and the second process executing the join continues to G.

```
m = 2
n = 3
A
FORK c
B
FORK e
FORK f
D
JOIN m,g; /* join 2 processes and goto label g */
g:G
    JOIN n,h; /* join 3 processes and goto label h */
h:H
    quit;
c:C
    JOIN m,g;
e:E
    JOIN n,h;
f:F
    JOIN n,h;
```

**FIGURE 6.16**
Generalized fork/join construct


Similarly, the three processes D, E, and F will each encounter the Join $n,h$ instruction. Each one of these three processes will decrement the counter $n$ (which is initialized to 3) once when it executes the Join instruction. Finally the process which reduces the counter $n$ to zero will continue to H. It is entirely upto the programmer to ensure that JOIN works in harmony with the FORK. As the example shows, JOIN can be used to join processes formed by any FORK. Since FORK can appear in loops and conditional statements, an arbitrary number of parallel processes can be created. Since JOIN can be used to join any of these processes to each other, a very powerful mechanism for starting and stopping parallel activities is provided by fork/join construct. A disadvantage is that there is a high potential that the resulting code may be unstructured, error-prone, and tough to understand.

## 6.5.2 Parbegin/Parend

Parbegin/parend construct is less powerful but more structured compared with Fork/Join. It is less powerful because it can only represent properly nested process flow graphs. That is, what parbegins must parend. An example follows:

---

**Example 6.8**
```
A
parbegin
     C
     begin
          B
          parbegin
               D
               E
               F
          parend
          G
     end
parend
H
```

This code consists of three modules to be executed sequentially: A, the module delimited by (outer) parbegin...parend and H. The second module consists of two sequential code modules: C and the module delimited by begin...end. The latter module in turn, consists of three modules: B, the module delimited by (inner) parbegin...parend and G.

During the execution of this code, processes D, E, and F are assigned to different processors and are executed independently. The parend statement makes sure all the processes that were started separately after the matching parbegin are completed before the next statement, G begins. Same is the case with the outer parbegin/parend pair which ensures that processes C and G end before H is executed.

---

The significant thing about parbegin/parend construct is that it identifies candidates for parallel execution so that the compiler can retain this simultaneity in the code. It does not force parallel execution. The degree of parallelism at run-time is dependent on the availability of processors at that time.

## 6.5.3 DOALL

Doall and constructs such as forall, pardo, doacross are like parbegin except that the statements eligible for parallel execution are in a loop. Thus the statements to be executed simultaneously are identified by an index rather than writing them out explicitly as was the case with parbegin/parend. Moreover, the number of statements between parbegin and parend is a predetermined, static quantity while the number of loop iterations in case of a doall loop may be computed during runtime and is thus a dynamic quantity.

### 6.5.4 Shared Variables

In shared-address programming, two types of variables are used: *shared* and *private* (or *local*). Typical declaration constructs are

```
shared float x;
private float z;
```

### 6.5.5 Send/Receive

The typical format for the SEND construct is

```
SEND(dest, type, length, flag, message)
```

where dest is the destination node address, type is the type of message, length is the message length, flag is a control parameter indicating the mode of message transmittal, and message is the actual message. There are two modes of message transmittal commonly used: *blocked* and *unblocked*. In the blocked mode, the sender is blocked until either the message received by the destination or a response is received. In the unblocked mode, the sender dispatches the message and continues its processing activity.

The typical format for the RECEIVE construct is

```
RECEIVE(source, type, length, flag, buf)
```

where source is the identity of the processor from which the message is received, type refers to one of several types of messages that can be received, length is the message length, flag is the blocking/nonblocking mode indicator, and buf is the memory buffer address where the message needs to be stored for further processing.

As in the case of vector processors and SIMD systems, the parallel programming support for MIMD systems is also provided both by special programming language constructs and compilers that transform the sequential code into parallel code. Refer to Chapter 5 for further details on the software.

## 6.6 Performance Evaluation and Scalability

As mentioned earlier, the granularity plays a very important role in determining the performance of an MIMD system. In a coarse grained application, the

*Multiprocessor Systems* 261

task size is large and thus the run time quantum ($R$) is quite large compared to the communication overhead ($C$). That is, the task in itself accomplishes a lot and not much communication is required between the various processes executing on different processors. Thus, the coarse-grained parallelism is characterized by a high $R/C$ ratio. On the other hand, a large number of small sized tasks running on many processors require frequent and may be long sessions of processor-to-processor communication. Such fine grain parallelism is characterized by a low $R/C$ ratio.

The definitions of Section 5.4 for the speedup, the efficiency, and the cost of parallel computer architectures apply to MIMD systems also and is illustrated by the following examples.

---

**Example 6.9**
Consider again the problem of accumulating $N$ numbers. The execution time on an SISD is of the $O(N)$. On an MIMD with $N$ processors and a ring interconnection network between the processors, the execution requires $(N-1)$ time units for communication and $(N-1)$ time units for addition. Thus the total time required is $2(N-1)$ or $O(2N)$, and hence the speedup is 0.5.

If the processors in the MIMD are interconnected by a hypercube network, this computation requires $\log_2 N$ communication steps and $\log_2 N$ additions, resulting in a total run time of $2\log_2 N$. Hence,

Speedup $S = N/(2\log_2 N)$ or $O(N/\log_2 N)$
Efficiency $E = 1/(2\log_2 N)$ or $O(1/\log_2 N)$, and
Cost $= N \times 2\log_2 N$ or $O(N\log_2 N)$

---

**Example 6.10**
The problem of accumulating $N$ numbers can be solved in two methods on an MIMD with $p$ processors and a hypercube interconnection network. Here, $p < N$ and we assume that $N/p$ is less than or equal to $p$. In the first method, each block of $p$ numbers are accumulated in $(2\log_2 p)$. Since there are $N/p$ such blocks, the execution time is $(2N/p \log_2 p)$. The resulting $N/p$ partial sums are the accumulated in $(2\log_2 p)$. Thus, the total run time is $(2N/p \log_2 p + 2\log_2 p)$. In the second method, each of the $p$ blocks of $N/p$ numbers is allocated to a processor. The run time for computing the partial sums is then $O(N/p)$. These partial sums are accumulated using the perfect shuffle network in $(2\log_2 p)$. Thus the total

run time is $(N/p + 2\log_2 p)$. The second method offers a better run time than the first.

If $N/p$ is greater than $p$, then in the first method above, further portioning of the computations to fit the $p$ processor structure is needed. Computation of run time for this case is left as an exercise. The run time characteristics for the second method would be the same as above for this case also.

### 6.6.1 Scalability

As defined earlier, the *scalability* of a parallel system is a measure of its ability to increase speedup as the number of processors increases. A parallel system is *scalable*, if its efficiency can be maintained at a fixed value, by increasing the number of processors as the *problem size* increases.

As the number of processors in an MIMD system increases, it should be possible to solve larger problems. For a given problem size, the efficiency of the system should be maximized while minimizing the run time. In general, it is possible to increase the number of processors and solve a larger problem (i.e., scale up the problem size). Depending on the problem domain and the parallel system architecture, a variety of constraints may be applicable. In *time-constrained scaling* (Gustafson, 1992), the problem size is scaled up with the increasing number of processors, while keeping the run time constant. In *memory-constrained scaling* (Sun and Ni, 1993), an attempt is made to solve the largest problem that can fit in the memory.

Let $T_s$ be the (best possible) run time of a problem on a single-processor system, and $T_p$ is the corresponding (best possible) run time on an MIMD system with $p$ processors. Then, the overhead cost $T_o$ is given by

$$T_o = T_p p - T_s \tag{6.2}$$

This overhead in MIMD systems is due to the interprocessor communication, load imbalance, extra computations required due to algorithms used, etc. The overhead is also a function of the problem size and the number of processors. From this equation, the parallel run time is

$$T_p = (T_o + T_s)/p \tag{6.3}$$

and hence the speedup $S$ is

$$S = T_s/T_p$$
$$= T_s p/(T_o + T_s) \tag{6.4}$$

The efficiency $E$ is

$$E = S/p$$
$$= T_s/(T_s + T_o)$$
$$= 1/(1 + T_o/T_s) \tag{6.5}$$

As $p$ increases, $T_o$ increases and if the problem size is kept constant, $T_s$ remains a constant. Then, from equation (6.4), $E$ decreases. If the problem size is increased while keeping $p$ constant, $T_s$ increases, $T_o$ grows at a smaller rate than the problem size (if the system is scalable) and hence $E$ increases. Thus, $E$ can be maintained at a fixed value, by increasing problem size as $p$ increases, for scalable systems.

The rate at which the problem size has to grow, as $p$ is increased to maintain a fixed $E$, is a characteristic of the application and the system architecture. The system is said to be *highly scalable* if the problem size needs to grow only linearly, as $p$ grows, to maintain a fixed efficiency.

From equation (6.5),

$$T_o/T_s = (1 - E)/E.$$

Hence,

$$T_s = ET_o/(1 - E)$$

For a given value of $E$, $E/(1 - E)$ is a constant $K$. Then,

$$T_s = KT_o \tag{6.6}$$

Since $T_o$ is a function of $p$, the problem size determined by $T_s$ above, is also a function of $p$. Equation (6.6) is called the *isoefficiency function* (Gupta and Kumar, 1993) of the MIMD system. A small isoefficiency function indicates that small increments in problem size are sufficient to maintain efficiency when $p$ is increased. A large isoefficiency function represents a poorly scalable system. If the parallel system is not scalable, the isoefficiency function does not exist.

Several other scalability criteria have been proposed over the years. (Refer to (Carmona and Rice, 1991; Kumar and Gupta, 1991; Marinescu and Rice, 1991; Tang and Li, 1990 for further details.)

### 6.6.2 Performance Models

Several performance models have been proposed over the years for evaluating MIMD systems. All these models are based on several assumptions that simplify the model. As such, none of the models may be a true representative of a practical MIMD system. Nonetheless, these models are helpful in understanding how MIMD systems behave and how their efficiency can be affected by various parameters. We now briefly describe three models introduced by Indurkhya, Stone, and XiCheng (Indurkhya et al., 1986).

### 6.6.2.1  The basic model

Consider a job that has been divided into $M$ tasks and that it has to execute as efficiently as possible on a multiprocessor with $N$ processors. The assumptions associated with this model are:

1. Each task is equal in size and takes $R$ time units to be executed on a processor.
2. If two tasks on different processors wish to communicate with each other they do so at a cost of $C$ time units. This is the communication overhead. However, if the two tasks are on the same processor (obviously executed one after the other) the overhead for their communication is zero.

Suppose $K_i$ tasks are assigned to the $i$th processor in the machine. Here, $i$ varies between 1 and $N$ and $K_i$ varies from 0 to $M$. Also assume that each processor was free before these tasks were distributed so that all the processors start executing at the same time. In this scenario, the processor that got the maximum number of tasks will take the longest to finish its run time quantum. In fact this will be the total run time of the job. Thus, the total run time of the job is $R(\max(K_i))$.

Here, $K_i$ tasks on the $i$th processor need to communicate with the rest of the $(M - K_i)$ tasks on the other processors. Only two tasks can communicate at the same time and each such communication introduces an overhead of $C$ time units. The total communication time for this job is

$$(C/2) \sum_{i=1}^{N} (K_i(M - K_i)) \qquad (6.7)$$

The total execution time (ET) for this job is thus

$$\text{ET} = R(\max(K_i)) + (C/2) \sum_{i=1}^{N} (K_i(M - K_i)) \qquad (6.8)$$

Now consider the following scenarios. First, all the tasks are assigned to a single processor. In this case, which is equivalent to an SISD, the first term of equation (6.8) becomes $RM$ since all the $M$ tasks will be executed serially by a single processor. The second term will be zero since all the tasks are executed on a single processor. Thus,

$$\text{ET} = RM \qquad (6.9)$$

In the second scenario, we assume that all the tasks are distributed equally among all the processors. Thus each processor gets $M/N$ tasks and hence $K_i = M/N$ for all $i$. Substituting this in equation (6.8):

$$ET = R(M/N) + CM_2/2 - CM^2/2N \qquad (6.10)$$

If we equate the results from equations (6.9) and (6.10) we get a point where the performance of an MIMD with equal task distribution among all the processors is the same as that of an SISD in which the single processor performs all the tasks. This threshold is represented by:

$$R/C = M/2 \qquad (6.11)$$

Thus if the task granularity is above the threshold $M/2$ then an equal distribution of tasks on MIMD systems is the best solution; but if it is below the threshold then executing the job on an SISD is most efficient, irrespective of the number of processors in the MIMD. This proves that the communication overhead should be kept within a certain limit in order to make parallel execution on an MIMD more beneficial than serial execution on an SISD.

The ratio of the right-hand sides of equations (6.9) and (6.10) gives the so-called speedup of an MIMD as compared to an SISD. Thus,

$$\text{Speedup} = RM/(RM/N + CM_2/2 - CM^2/2N) \qquad (6.12)$$

From this equation it is clear that if $N$, the number of processors is large, speedup is a constant $R/CM$ for a given granularity and a given number of tasks. Thus as $N$ increases, the speedup reaches saturation. In other words, simply increasing the number of processors to improve the performance of an MIMD is not an intelligent step because after a certain stage, the speedup stops increasing while the cost keeps going up.

### 6.6.2.2 Model with linear communication overhead

In the previous model we have considered a case in which each task has to undergo a unique communication with each task on other processors. This results in quadratic communication overhead. However, if we assume that each task has to communicate with all other tasks but the information contents are the same for all the other tasks, the communication overhead becomes linear with respect to $N$ since the given task does not have to uniquely call the other tasks one by one. With this model the cost of an assignment of a job with $M$ tasks to an MIMD with $N$ processors is

$$ET = R(\max(K_i)) + CN \qquad (6.13)$$

Considering an equal distribution of tasks among all the processors, this equation can be written as

$$ET = RM/N + CN \qquad (6.14)$$

Thus we see that as $N$ increases the second term increases linearly with $N$ while the first decreases with $N$. Thus initially as $N$ increases, the execution time decreases until it reaches a minima after which the addition of a processor to the system has degradable effect rather than a beneficial one. From equation (6.9) the execution time decrease with the addition of a new processor is:

$$\text{ET decrease} = RM/N + CN - RM/N + 1 - C(N+1)$$
$$= RM/N(N+1) - C \qquad (6.15)$$

At the point of minima discussed above the performance of the system is the best. That is, the right-hand side of equation (6.10) is zero at such a point. Hence,

$$R/C = N(N+1)/M$$

and hence,

$$N = \sqrt{(MR/C)} \qquad (6.16)$$

Thus, with this model the magnitude of parallelism is only the square root of what was anticipated.

### 6.6.2.3  A model with overlapped communication

In all the models discussed so far, we have considered useful computation and communication overhead as two entities that are exclusive of each other, that is, they cannot be performed concurrently. This need not be the case in practice. In this model we assume that the communication time fully overlaps with the useful computation being performed by the other processors.

With this model the total execution time, ET will be the greater of the two terms which represent the run time and the communication overhead, respectively. Again with $K_i$ tasks having been allocated to the $i$th processor, the total run time for this set up will be $R1 = R(\max(K_i))$. The total overhead due to communication will be, as in the first model

$$C1 = C/2 \sum_{i=1}^{N}(K_i(M - K_i))$$

Then,

$$ET = \max(R1, C1) \qquad (6.17)$$

For full utilization of a fully overlapped model it is desirable that $R1 = C1$. That is, the communication overhead requires the same amount of time as the useful computation. If this is also combined with the condition of equal task distribution among all the processors, we get

$$R1 = RM/N \quad \text{and} \quad C1 = CM^2/2(1 - 1/N)$$

and equating them we get

$$R/C = NM/2 \qquad (6.18)$$

or

$$N = 2R/CM$$

This equation shows that it is advisable to decrease $N$ as the parallelism (i.e., $M$) increases. Also since $C1$ grows $M$ times faster than $R1$ it is really important to keep it less than $R1$. If $C1$ is greater than $R1$, the overall execution time becomes solely dependent on the communication overhead.

### 6.6.2.4 A stochastic model

In practice all the tasks are not of equal length. Then the objective is to scatter all the tasks on the processors such that all the processors are busy for equal periods of time. In the previous models, this objective was fulfilled by assigning equal number of tasks to all the processors. In the stochastic model it is clear that equal distribution (number wise) of tasks will not lead to equal processing times for all the processors. If the workload is uneven, that is unequal number of tasks are to be assigned to different processors, it may be possible to assign tasks to processors in such a way that overhead is greatly reduced. As is clear from equation (6.8), the communication overhead increases if the distribution of tasks is made more uniform over all the processors. In other words, the more uneven the distribution, the lesser is the communication cost. Moreover having unequal size tasks favor unequal distribution of tasks so that the average busy time of each processor remains roughly the same.

In summary, the run time of a program diminishes as the number of processors go up, but the overhead cost tends to shoot up with increasing number of processors. The rate of increase of overhead cost may be much faster than the rate of decrease of the run time. Also, the task granularity plays a crucial part in determining the efficiency of a given model.

Several other analytical techniques have been used in performance evaluation. The major shortcoming of these techniques is that they do not fully represent the operating scenario of the system, since each technique makes several simplifying assumptions to manage the complexity of the model. Use of benchmarks is considered a more practical method, for performance evaluation.

Chapter 2 provided the details of representative benchmarks. In 1985, the National Institute for Standards and Technology (NIST) held a workshop on techniques for performance evaluation of parallel architectures and has published a list of benchmarks to include: Linpack, Whetstones, Dhrystones, NAS Kernel, Lawrence Livermore Loops, Fermi, JRR, and Mendez. Levitan (1987) has proposed a suite of synthetic benchmarks for evaluation of interconnection structures. The PERFECT and SLALOM benchmarks are more recent ones. Since the amount of parallelism varies among benchmarks, each benchmark will provide a different performance rating for the same architecture. Also, since the ability of architecture to utilize the parallelism in the benchmark varies, the rating provided by a given benchmark varies across the architectures.

## 6.7 Example Systems

This section provides a brief description of three commercial multiprocessor systems: Thinking Machine Corporation's CM-5, a hybrid SIMD/MIMD system; Cray Research Corporation's T3D, a distributed memory, shared-address-space MIMD system; and IBM System X. The first two systems are no longer in production.

### 6.7.1 Thinking Machine Corporation's CM-5

The Thinking Machine Corporation's Connection Machine 5 (CM-5) is a hybrid MIMD/SIMD multiprocessor system. The number of processors can be scaled from 32 to 16,384 providing an observed peak performance of 20 GFLOPS. Optionally, four vector units can be added to each processor, resulting in a peak performance of 2 teraflops. The CM-5 is controlled by one or more front-end workstations (Sun Microsystems' SPARC 2, Sun-4, or Sun-600), which execute the CM-5 operating system, CMost. These front-end workstations (control processors) are dedicated to either PE array partition management or I/O control, and control all the instructions executed by the CM-5.

Figure 6.17 shows the system block diagram of the CM-5. There are three independent network subsystems — the Data Network, the Control Network, and the Diagnostic Network. All system resources are connected to all these networks. Processing nodes are not interrupted by network activity, which permits message routing and computational tasks to be performed in parallel. The CP broadcasts a single program to all the nodes. Each node runs this program at its own rate, utilizing the control network for synchronization. The library primitives provided with the machine, allow the implementation of synchronous communications for data parallel parts and asynchronous communications for the message-passing parts of the program.

The Data Network provides for data communications between CM-5 components. It has a Fat tree topology. Each set of four processing or storage nodes has a network switch connecting them together. This switch provides for the data communication between the four nodes and to the upper layers of

*Multiprocessor Systems* 269

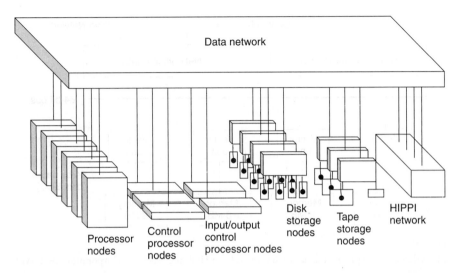

**FIGURE 6.17**
CM-5 system block diagram (Reprinted from Thinking Machine Corporation, 1991. With permission.)

the network. Four of these groups of processing or storage nodes are in turn connected by a higher-level switch. Four of these larger groups are in turn similarly connected, to provide an aggregate bandwidth of 1280 MB/sec. At the bottom layer, each node is connected to its corresponding network switch by two bi-directional links of 5 MB/sec capacity.

The Control Network subsystem coordinates the processor interactions required to implement inter-processor broadcasting and synchronization. It also provides the interface between the CM-5 front end processors and the corresponding processor nodes, and provides protection against multiuser interference.

The Diagnostic Network subsystem provides the interfaces required to perform internal system testing of device connectivity and circuit integrity. It can also access machine environmental variables (temperature, air flow, etc.) for detection of failure conditions.

Each processing node consists of a 22-MIPS SPARC microprocessor, 32 MB of memory, four 32-MFLOP vector processing units (64-bit operands) each with 8 MB memory, and a network interface (Figure 6.18). The vector units implement a full floating-point instruction set with fast divide and square root functions. The SPARC is the control and scalar processing resource of the vector units. It also implements the operating system environment and manages communication with other system components via the network interface.

The Scalable Disk Array is an integrated and expandable disk system that provides a range of 9 GB to 3.2 TB of storage. Data transfers between CM-5 and the Disk Array can be sustained at 12 MB to 4.2 GB $\text{sec}^{-1}$. The Disk Array is an array of disk storage nodes. Each node contains: a network interface

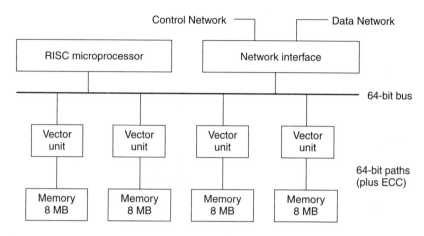

**FIGURE 6.18**
Block diagram of a processing node (Reprinted from Thinking Machines Corporation, 1991. With permission.)

implemented by an RISC microprocessor controller, disk buffer, four SCSI controllers, and eight disk drives. The Scalable Disk Array features are implemented via the vendor's CM-5 Scalable File System (SFS) software, which executes on an SPARC-based I/O Control Processor.

The I/O nodes are the third class of computational resources in a CM-5 system. These nodes include magnetic tape and network communications facilities. The integrated tape subsystem is implemented via a specialized I/O node which has local buffer memory and a pair of SCSI-2 channel controllers that can connect to a maximum of seven devices per controller. Communication with other computers is provided by the HIPPI I/O node. Serial and parallel TCP/IP and user HIPPI-FP style interfacing through sockets is provided by the HIPPI subsystem. CM-5 also supports Ethernet and FDDI communications.

CMost provides the integrated user interface in the form of the X-Window System and OSF/Motif. The software environment integrates the debugging, performance analysis, and data visualization facilities into the integrated user interface. The following languages are available: C*, CM FORTRAN, and *Lisp, which are respective supersets of C, FORTRAN-90, and Lisp (for details refer to Chapter 5).

### 6.7.2 Cray Research Corporation's T3D

The Cray T3D system (Kessler and Schwarzmeier, 1993; Koeninger et al., 1994) is the first massively parallel processor (MPP) from Cray Research Incorporated. It integrates industry-leading microprocessors with a Cray-designed system interconnect and high-speed synchronization mechanisms to produce a balanced scalable MPP.

*Multiprocessor Systems* 271

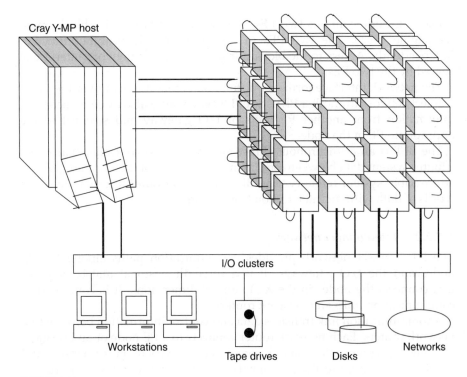

**FIGURE 6.19**
Cray T3D system (Koeninger et al., 1994) (Reprinted from *Digital Technical Journal*. With permission.)

The MPP section of the T3D system (Figure 6.19) contains hundreds or thousands of DEC Alpha 21064 microprocessors, each associated with its own local memory. It is connected to a parallel vector processor (PVP) host computer system (either Cray Y-MP or Cray C90). All applications written for the T3D system are compiled on the host. The scalar and vector oriented parts of the application run on the host while the parallel portions run on the MPP. The MPP consists of three components: Processing element nodes, Interconnect network, and I/O Gateways.

### 6.7.2.1 Processing element nodes

Each processing element (PE) contains a microprocessor, local memory, and the support circuitry. The local memory is 16 or 64 MB of DRAM with a latency of 87 to 253 nsec and a bandwidth of up to 320 MB sec$^{-1}$. Remote memory is directly addressable by the processor with a latency of 1 to 2 $\mu$sec and a bandwidth of over 100 MB sec$^{-1}$. The total size of memory in a T3D system is the number of PE times the size of each PE's local memory. The support circuitry extends the control and addressing functions of the microprocessor and facilitates data transfers to or from local memory.

Each PE node contains two PEs, a network interface and a block transfer engine (BLT). The two PEs in the node are identical but function independently. Access to the block transfer engine and network interface is shared by the two PEs.

The network interface formats information before it is sent over the interconnect network to another PE node or I/O gateway. It also receives incoming information from another PE node or I/O gateway and steers it to the appropriate PE in the node.

The BLT is an asynchronous direct memory access controller that redistributes system data between local memory in either of the PEs and globally addressable system memory. The BLT can redistribute up to 65,536 64-bit words of data without interruption from the PE.

### 6.7.2.2  Interconnect network

The interconnect network provides communication paths among the PE nodes and the I/O gateways. It is a three-dimensional matrix of paths that connects the nodes in the $X$, $Y$, and $Z$ dimensions. It is composed of communication links and network routers.

A communication link transfers data and control information between two network routers. Each network router connects to a PE node or an I/O gateway node. A communication link connects two nodes in one dimension and consists of two unidirectional channels. Each channel contains data, control, and acknowledge signals.

The interconnect network is a bi-directional torus in which a link connects the smallest numbered node in a dimension directly to the largest numbered node in the same dimension. This type of connection forms a ring where information can transfer from one node, through all of the nodes in the same dimension, and back to the original node.

The network uses the dimension-order routing. That is, when the message leaves a node, it travels through the network in the $X$ dimension first, then through the $Y$ dimension, and finally through the $Z$ dimension, to the destination node.

### 6.7.2.3  I/O gateways

The I/O gateways transfer system data and control information between the host system and the MPP or between the MPP and an input/output cluster. An I/O gateway can transfer information to any PE in the interconnect network.

The T3D system provides three mechanisms for hiding the start up time (latency) of remote references: Prefetch Queue, Remote Processor Store, and Block Transfer Engine.

### 6.7.2.4  Prefetch queue

The DEC alpha instruction set contains an opcode (FETCH) that permits a compiler to hint the hardware of an upcoming memory activity. The T3D shell hardware uses FETCH to initiate a single word remote memory read that will fill a slot that is reserved by the hardware in the external prefetch queue. The prefetch queue acts as an external memory pipeline. As the processor issues each FETCH instruction the shell hardware reserves a location in the queue for the return data and sends a memory read request packet to the remote node. When the read data returns to the requesting processor, the shell hardware writes the data to the reserved slot in the queue. The processor retrieves the data from the queue by executing a load instruction from a memory mapped register that represents the head of the queue. The data prefetch queue can store up to 16 words, so the processor can issue up to 16 FETCH instructions before executing any load instruction to pop the data from the head of the queue.

### 6.7.2.5 Remote processor store

The Alpha processor stores to remote memories need not wait for the completion of a single store operation. This is an effective communication mechanism when the producer of the data knows which PEs will immediately need to use the data. The Alpha processor has four 4-word write buffers on chip that try to accumulate four words (a cache line) of data before performing the actual external store. This feature increases the effective bandwidth. In the T3D system, a counter is set up in the PE shell circuitry. Each time the processor issues a remote store, the counter is incremented and each time a write operation completes, the counter is decremented. The processor can read the counter and determine when all the write operations are completed.

### 6.7.2.6 Block transfer engine

The block transfer engine (BLT) is an asynchronous direct memory access controller used to redistribute data between remote and local memory. The BLT operates independently of the processors at a node. The processor initiates BLT activity by storing individual request information in the memory-mapped control registers. The overhead is significant in this setup work and hence, the BLT is most effective for large data block moves.

The T3D system provides four hardware synchronization primitives: Barrier, Fetch and increment, Lightweight message, and Atomic swap. These facilitate synchronization at various levels of granularity and support both control parallelism and data parallelism.

In summary, the major characteristics of the T3D system are

1. High-speed microprocessors
2. Distributed shared memory
3. High-speed interprocessor communications
4. Latency hiding
5. Fast synchronization
6. Fast I/O capabilities

The T3D system is scalable from 32 to 2048 processors, from 4.2 to over 300 GFLOPS peak performance, and from 512 MB to 128 GB of memory.

#### 6.7.2.7  Programming model

The system uses CRAFT, a FORTRAN programming model with implicit remote addressing. It provides a standard, high-level interface to the hardware, both the traditional data parallel and message-passing MPP programming capabilities, and the work sharing capability between the host and the MPP. Parallelism can be expressed explicitly through message-passing library calls or implicitly through data parallel constructs and directives, taking advantage of the compiler to distribute data and work.

#### 6.7.2.8  Operating system

The UNICOS MAX operating system was designed to reuse Cray's UNICOS operating system, which is a superset of standard UNIX operating system. UNICOS MAX is a multiuser operating system which provides a powerful supercomputing environment that enhances application development, system interoperability, and user productivity.

UNICOS MAX is a distributed operating system wherein functions are divided between the MPP and the UNICOS operating system on the host. The bulk of the operating system runs on the hosts with only microkernels such as application data storage and processing running on MPP processors.

#### 6.7.2.9  Performance

A 256-processor T3D system was the fastest MPP running NAS parallel benchmarks. Although the Cray C916 PVP system ran six of the eight benchmarks faster than the T3D, the price/performance ratio is better for T3D, since the cost of C916 is about three times that of T3D. It is expected that a 512-processor T3D will outperform C916 in six of the eight benchmarks.

### 6.7.3  IBM System X

System X introduced in 2003, is based on an Apple G5 platform with the IBM PowerPC 970 64-bit processors. It uses Gigabit Ethernet switches for communication (Mellanox' InfiniBand semiconductor technology) and is cooled by a hybrid liquid–air cooling system. It uses the *Mac OS X operating platform*. Language support includes C, C++ optimizing compilers (IBM xlc and gcc 3.3), and Fortran 95/90/77 compilers (IBM xlf and Nag ware).

The Apple G5 processor has two double-precision floating-point units. Each is capable of a fused, multiple-add operation per cycle to get 2 flops per cycle. This implies that 2 GHz corresponds to 8 Giga floating-point operations per second (GFlops). Thus each dual G5 can give a maximum of 16 GFlops of double precision performance. The G5 has a superscalar pipelined execution

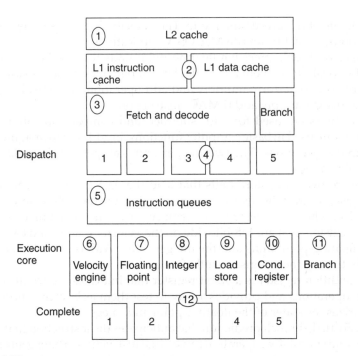

**FIGURE 6.20**
The power PC G5 architecture (Courtesy of Apple Computers.)

core and uses 128 bit, 162 instruction SIMD unit. The Power PC G5 processor architecture is shown in Figure 6.20.

The L2 cache is 512K with 64-MBps access to data and instructions. The L1 Cache is configured into two caches: Instruction cache and data cache. Instructions are prefetched from the L2 cache into the direct-mapped 64K L1 cache at 64GBps. L1 data cache is 32K and can prefetch up to eight active data streams concurrently.

Up to eight instructions per clock cycle are fetched from the L2 cache and decoded and divided into smaller operations. Instructions are sent into the execution core and data is loaded into the large number of registers following the functional units.

Before dispatching, the instructions are arranged into groups of up to 5. The PowerPC G5 tracks up to 20 groups at a time or 100 (5 ∗ 20) individual instructions and manages 20 instructions in each of the five dispatch groups and additional 100-plus instructions in the various fetch, decode, and queue stages.

Each functional unit has its own queue where multiple instructions are stored. The instruction group that is dispatched is broken into discrete instructions that go to the appropriate functional unit.

The PowerPC G5 uses a dual-pipelined Velocity Engine with two independent queues and dedicated 128-bit registers and data paths for instruction and dataflow. This vector-processing unit speeds up data manipulation by applying the SIMD processing mode. The Velocity Engine uses the set of 162 instructions, so that it can run Mac OS X applications.

It uses two double-precision floating-point units that let the G5 complete at least two 64-bit mathematical calculations per clock cycle. Floating-point units are used to perform multiply and add operations and are also capable of doing fused multiply add (FMAC) instruction.

Integer units are used for add, subtract, and compare operations. Such operations are used in basic computer functions, in addition to imaging, video and audio applications. The two integer units can perform both 32-bit and 64-bit calculations.

There are two load/store units that load the data from L1 cache into the data registers. After the instructions manipulate the data, these units store it back to L1 cache, L2 cache, or main memory. Each functional unit is equipped with 32 registers that are 128-bit wide on the Velocity Engine and 64-bit wide on the floating-point units and the integer units. With two load/store units, it is able to keep these registers full of data.

The condition register is a 32-bit register that indicates the results of comparison operations and provides a way to test them as branch conditions. It summarizes the states of the floating-point and integer units.

The G5 can do branch prediction that anticipates the instruction that should go next, and speculative operation causes that instruction to be executed. If the prediction is correct, the processor works efficiently — since the speculative operation has executed an instruction before it is required. If the prediction is incorrect, the processor clears the redundant instruction and associated data, resulting in an empty space called a *pipeline bubble*. Pipeline bubbles reduce performance as the processor marks time waiting for the next instruction.

Once operations on the data are complete, the instructions are reconstructed into the original groups of five and the load/store units store the data in cache or main memory for further processing.

The System X uses InfiniBand's card. The InfiniBand network uses a switched, point-to-point fabric employing virtual channels for communication which has two ports on each node connecting into the network at 20 GBps full-duplex bandwidth. Each node has a connection open to each other node and can hold up to 150 K connections. This translates into a latency of less than 10 $\mu$sec. All cluster nodes are interconnected by InfiniBand links through 24 96-Port InfiniBand switches. InfiniBand combines storage area networks and system area networks. InfiniBand interfaces bypass the bus and communicate directly with the memory controller.

## 6.8 Summary

The basic architectural models for MIMD systems were introduced in this chapter. Problems involved in designing these systems, concepts of programming such systems, and performance evaluation were also discussed. Details of three commercial multiprocessor systems were provided.

The field of multiprocessor systems has been a dynamic one recently in the sense that several experimental and commercial MIMD systems have been built, brought to market, and discarded. The hardware technology of today allows the development of large MIMD systems. But, tailoring the applications to take advantage of the parallelism in hardware is still nontrivial. Parallel programming languages and corresponding compilers and software development support environments are still evolving.

Some of the problems with MIMD system application are:

1. Selecting the proper model of parallel computation.
2. Parallel programming languages that allow easy representation of: task creation, deletion, and synchronization and send/receive message-passing primitives.
3. Reduction of overhead by: overcoming data dependencies of tasks, optimal allocation of processors, minimizing contention for resources and idle waiting, and optimal scheduling of tasks.
4. Scalability.
5. The control of software development and operational control of the MIMD system is also complex. A primary concern is how the development of application software for a multinode computing system should be approached. Much work has been done concerning the need for maintaining control of software developed for SISD computers. With the $N$-fold complexity of concurrent architectures, the concern of software development control becomes paramount. More sophisticated software development environments are needed if a multitude of programs are to be developed within reasonable cost and time constraints. Williams (1990), Brawer (1989), and Perrott (1987) provide further details on these topics.

Obviously, the single-processor architecture is the most efficient solution, if the performance requirements of the application can be met. If not, an SIMD solution might be evaluated. If the application does not suit the SIMD solution, the MIMD solution would be appropriate. In addition to the performance and cost constraints, the application characteristics dictate the selection of architecture.

## Problems

6.1. It is required to design a general-purpose multiprocessor system using processor and memory elements. Identify the minimum set of characteristics that each element needs to satisfy.

6.2. Four 16-bit processors are connected to four 64K × 16 memory banks through a crossbar network. The processors can access 64K memory directly. Derive the characteristics of the crossbar network to efficiently run this multiprocessor system. Show the hardware details. Describe the memory mapping needed. How is it done?

6.3. Study any multiprocessor system you have access to, to answer the following:

1. What constitutes a 'task'?
2. What is the minimum task switching time?
3. What synchronization primitives are implemented?

6.4. The fetch-and-add can be generalized to fetch-and-op($a, b$), where $a$ is a variable and $b$ is an expression. This operation returns the value of $a$ and replaces it with ($a$ op $b$). What would be the advantages of this operation over the synchronization primitives described in this chapter?

6.5. Analyze the matrix multiplication algorithm of this chapter to derive the optimum partitioning and storage of matrices **A** and **B** so that all the spawned tasks can run simultaneously.

6.6. Consider the following computation:

$$c_i = a_i * b_i + c_i * d_i$$

where $i = 1$ to $N$. Write a high-level program for this computation using (1) fork/join (2) parbegin/parend, and (3) doall primitives.

6.7. Estimate the computation time for the program of (1) in the above problem assuming the following time characteristics:

| Operation | Execution time |
|---|---|
| Addition | 1 |
| Multiplication | 3 |
| Fork | 10 |
| Join | 4 |

Assume that the tasks start execution as soon as they are spawned and there is no waiting time.

6.8. What is the ideal task granularity in Problem (6.7) (i.e., when will the computation time be a minimum) if the number of processors is 4?

6.9. An algorithm requires access to each row of an $N \times N$ matrix. Show the storage of matrix to minimize the access time if the multiprocessor consists of $N$ processors and $N$ memory banks interconnected by (a) crossbar (b) bus.

6.10. Repeat Problem (6.9), if the algorithm accesses both rows and columns of the matrix.

6.11. Two concurrent processes X and Y each execute the following sequence of statements:

```
WAIT (A)
WAIT (B)
SIGNAL (B)
SIGNAL (A)
```

where A and B are binary semaphores.

1. Can these processes deadlock? If so, when?
2. If X remains the same, but Y is changed to:

```
WAIT (B)
WAIT (A)
SIGNAL (A)
SIGNAL (B)
```

How does the deadlock condition change?

6.12. For the following program:

```
for i = 1, n
  for j = 1, m
    c[i] = c[i] + a[i,j] * b[i,j]
  endfor
endfor
```

1. Determine the dependencies among the statements.
2. Interchange $i$ and $j$ loops and determine dependencies among the statements.
3. Which of the above two programs would you prefer to achieve each of the following implementation goals: vectorization, shared-memory MIMD, message-passing MIMD?

6.13. Consider an MIMD with $N$ processors and $N$ memories. It is required to compute the row sum of an $N \times N$ matrix using this system. Show the storage format of the matrix among the $N$ memories and the corresponding time to compute the row-sum, if the following interconnection networks are used:

1. Crossbar
2. Single-bus

6.14. Study the PCI and USB bus architectures. What support do they provide for configuring multiprocessor systems?

6.15. Assume that the synchronization and communication overhead in an $N$ processor system is $10N$ instruction cycles. Plot the Speedup offered by this system as a function of the number of processors, if synchronization and communication is required every 100, 1000, and 10,000 instructions.

## References

Adam, T., Chandy, K., and Dickson, J. A. Comparison of list schedulers for parallel processing systems. *Communication of ACM*, 17, 1974, 685–690.

Agrawal, D. P., Janakiram, K. V., and Pathak, G. C. Evaluating the performance of multicomputer configurations. *IEEE Computer*, 19(5) May 1986, 23–37.

Ajmone, M. M., Balbo, G., and Conte, G. *Performance Models for Multiprocessor Systems*. New York: McGraw-Hill, 1989.

Alamasi, G. S. and Gottlieb A. *Highly Parallel Computing*. Menlo Park, CA: Benjamin/Cummings, 1989.

Anderson, T. The performance of spin lock alternatives for shared-memory multiprocessors. *IEEE Transactions on Parallel Distributed Systems*, 1, 1990, 6–16.

Annaratone, M. and Ruhl, R. Performance measurements on a commercial multiprocessors running parallel code. In *Proceedings of the 16th Annual IEEE International Symposium on Computer Architecture*, pp. 307–314, 1989.

Apple Computers, *Power PC G5 Processor*. Available at http://www.apple.com

Archibald, J. and Baer, J. L. Cache coherence protocols: evaluation using a multiprocessor simulation model. *ACM Transactions on Computer Systems*, 4, 1986, 273–298.

Bashir, A., Susarla, G., and Karavan, K. A statistical study of a task scheduling algorithm. *IEEE Transactions on Computers*, C-32, 1983, 774–777.

Bawerjee, P. et al. Algorithm based fault tolerance on a hypercube multiprocessor. *IEEE Transactions on Computers*, 39, 1990, pp. 124–132.

Benes, V. E. *Mathematical Theory of Communication Networks and Telephone Traffic*. New York: Academic Press, 1965.

Benes, V. E. On rearrangeable three-stage connecting networks. *The Bell System Technical Journal*, 4, 1962, 1481–1492.

Bhuyan, L. N. and Agrawal, D. P. Generalized hypercube and hyperbus structures for a computer network. *IEEE Transactions on Computers*, C-33, 1984, pp. 184–197.

Brawer, S. *Introduction to Parallel Programming*, New York: Academic Press, 1989.

Carmona, E. A. and Rice, M. D. Modeling the serial and parallel fractions of a parallel algorithm. *Journal of Parallel and Distributed Computing*, 5, 1991, 76–83.

Censier, L. M. and Feautrier, C. A new solution to coherence problems in multicache systems. *IEEE Transaction on Computers*, C-27, 1978, 77–87.

Clos, C. A study of nonblocking switching networks. *The Bell System Technical Journal*, 1953, 406–424.

Deital, H. M. *Operating Systems*. Reading, MA: Addison-Wesley, 1990.

Dinning, A. A survey of synchronization methods for parallel computers. *IEEE Computer*, 22, 1989, 66–77.

Dijkstra, E. W. Cooperating Sequential Processes. In F. Genuys (ed.), Programming Languages, pp. 43–112, New York, NY: Academic Press, 1968.

Eggers, J. S. and Katz, R. H. evaluating the performance of four snooping cache coherency protocols. In *Proceedings of the 16th Annual IEEE International Symposium on Architecture*, 17, pp. 3–14, 1988.

Eisner, C., Hoover, R., Nation, W., Nelson, K., Shitsevalov, I., and Valk, K. A methodology for formal design of hardware control with application to cache coherence protocols. In *Proceedings of the 37th Conference on Design Automation*, pp. 724–729, 2000.

El-Rewini, H. and Lewis, T. G. Scheduling parallel program tasks onto arbitrary target machines. *Journal of Parallel and Distributed Computing*, 4, June 1990, 126–133.

Feng, T-Y. A survey of interconnection networks. *IEEE Computer*, 14, 1981, 12–27.

Gallant, J. Cache coherency protocols. *EDN*, 1991, 2(4), 41–50.

Ghezzi, C. and Jazayeri, M. *Programming Language Concepts*. New York: John Wiley & Sons, 1982.

Goke, G. and Lipovsky G.J. Banyon networks for partitioning multiprocessor systems. In *Proceedings of the First Annual Symposium on Computer Architecture*, pp. 21–28, December 1983.

Goodman, J. R. Using Cache memory to reduce processor memory traffic. In *Proceedings of the 10th International Symposium on Computer Architecture*, June 1983, 102–107.

Graunke, R. and Thakkar, S. Synchronization algorithms for shared-memory multiprocessors. *IEEE Computer*, 23, 1990, 175–189.

Gupta, A. and Kumar, V. Performance properties of large scale parallel systems. *Journal of Parallel and Distributed Computing*, 19, 7, 1993, 252–261.

Gustafson, J. L. The consequences of fixed time performance measurements. In *Proceedings of the 25th Hawaii International Conference on System Sciences*, Vol. III, pp. 113–124, 1992.

Hagersten, E., Landin, A., and Haridi, S. DDM — A Cache-only memory architecture. *IEEE Computer*, 25, 1992, 44–54.

Indurkhya, B., Stone, H. S., and XiCheng, L. Optimal partitioning of randomly generated distributed programs. *IEEE Transactions on Software Engineering*, SE-12, 1986, 483–495.

Kallstrom, M. and Thakkar, S. S. Programming three parallel computers. *IEEE Software*, 5, 1988, 11–22.

Kessler, R. and Schwarzmeier, J. Cray T3D: A new dimension for cray research. In *Proceedings of COMPCON*, pp. 176–182, 1993.

Kim, D., Chaudhuri, M., and Heinrich, M. Leveraging Cache coherence in active memory systems. In *Proceedings of the 16th International Conference on Supercomputing*, pp. 2–13, 2002.

Koeninger, R. K., Furtney, M., and Walker, M. A shared memory mpp from cray research. *Digital Technical Journal*, 6, 8–21, 1994.

Knuth, D. E. An empirical study of Fortran programs. *Software Practice and Experiences*, 1971, 105–133.

Kumar, V. and Gupta, A. Analyzing Scalability of Parallel Algorithms and Architectures. *Technical Report 91–18*, Computer Science Department, University of Minnesota, 1991.

Levitan, S. P., Measuring communication structures in parallel architectures and algorithms. *Characteristics of Parallel Programming*, Cambridge, MA: MIT Press, 1987.

Marathe, J., Nagaranjan, A., and Mueller, F. Detailed Cache Coherence characterization for openMP benchmarks. In *Proceedings of the 18th Annual International Conference on Supercomputing*, pp. 287–297, 2004.

Marinescu, D. C. and Rice, J. R. On High Level Characterization of Parallelism, *Technical Report CSD-TR-1011*, Computer Science Department, Purdue University, 1991.

Mellanox Technologies, InfiniBand Architecture Debuts #3 Supercomputer on the Planet, 2003, http://www.mellanox.com/presentations.html.

Papamarcos, M. and Patel, J. A low overhead coherence solution for multiprocessors with private cache memories. In *Proceedings of the 11th International Symposium on Computer Architecture*, pp. 348–354, 1984.

Perrott, R. H. *Parallel Programming*. New York: Addison-Wesley, 1987.

Pfneiszl, H. and Kotsis, G. *Benchmarking Parallel Processing Systems — A Survey*. Vienna, Austria: University of Vienna, Institute of Applied Computer Science, 1996.

Rattner, J. Parallel supercomputers tackle tough research questions. *R&D Magazine*, July 1992, 48–52.

Shiva, S. G. *Computer Design and Architecture*. New York: Marcel Dekker, 2000.

Siegal, H. J. *Interconnection Networks for Large-Scale Parallel Processing*, Lexington, MA, Lexington Books, 1985.

Silberschatz, A., Galvin, P., and Gragne, G. *Operating System Concepts*. New York: Addison Wesley, 2003.

Skillicorn David, B. and Talia, D. *Programming Languages for Parallel Processing*. Washington: IEEE Computer Society Press, 1996.

Stallings, W. *Operating Systems*, 4th ed., Prentice Hall, 2002.

Stone, H. S. Parallel processing with the perfect shuffle. *IEEE Transactions on Computers*, C-20, 1971, 153–161.

Sun, X. H. and Ni, L. M. Scalable Problems and Memory-Bounded Speedup. *Journal of Parallel and Distributed Computing*, 19, 1993, 27–37.

Tang, C. K. Cache system design in a tightly coupled multiprocessor system. In *Proceedings of the AFIP National Computer Conference*, 1976.

Tang, Z. and Li, G. J. Optimal granularity of grid iteration problems. In *Proceedings of the International Conference on Parallel Processing*, I, 1990, 111–118.

Tanenbaum, A. *Modern Operating Systems*. Upper Saddle River, NJ: Prentice-Hall, 2001.

Thinking Machines Corporation. *CM-5 Desktops-to-TeraOps Supercomputing*. Cambridge, MA, 1993.

Thinking Machines Corporation, *CM-5 Product Specifications*, Cambridge, MA, 1991.

Williams, S. A. *Programming Models for Parallel Systems*. New York, NY: John Wiley & Sons, 1990.

Wilson, R. CPUs work in multiples. *EE Times*, January 24, 2005, pp. 41–43.

Wu, C.-L. and Feng, T.-Y. On a class of multistage interconnection networks. *IEEE Transactions on Computers*, C-29, 1980, 694–702.

Yeh, P. C., Patel, J. H., and Davidson, E. S. Shared Cache for multiple stream computer systems. *IEEE Transactions on Computers*, C-32, January 1983, 32–39.

# 7
## Current Directions

The primary goal of innovative architectures described in this book, is either a dramatic improvement in the cost/performance ratio compared with existing designs or a highly scalable performance along with a good cost/performance ratio. In general, there are two approaches to achieving the highest performance with the least cost. The first one is an *evolutionary* approach that attempts to improve the performance of the single-processor system to its possible limit (the topic of Chapters 1, 3, and 4), and the second is the *revolutionary* approach that uses a large number of processors and minimizes the overhead involved, to achieve linear speedup (the topic of Chapters 5 and 6).

The first approach is mainly a result of advances in hardware technology yielding powerful processors at a very low cost. The advantage of this approach is that the applications and algorithms remain mostly unchanged, although changes to compilers and operating system may be needed to accommodate the new hardware. As and when the new hardware is available there will be a market for it, since the users tend to update their systems with the new ones in an incremental fashion. For instance, the early versions of the Cray series of supercomputers have followed this strategy, whereby the number of processors in the system is kept relatively low (up to 16 in the Cray Y-MP series) while the processors are made increasingly powerful in terms of their performance.

The advantage of the second approach is that the hardware (processors, memories, and interconnection networks) can be simply replicated. But, this massively parallel-processing (MPP) strategy requires a rethinking of application algorithms and data structures to effectively utilize the multiprocessor architecture. There have been successful efforts to device languages that represent parallelism explicitly, compilers that retain the parallelism expressed by the code and operating systems to manage the multiprocessor system resources efficiently. Even with the availability of these, the redesign of applications is not trivial.

Several multiprocessor architectures have been proposed over the years, some for general-purpose applications and some for special-purpose applications. For the general-purpose designs to survive, the development cost should be recuperated by the sale of a large number of systems, while the special-purpose application should justify the cost of architecture development. Nevertheless, new and more powerful architectures are being introduced at a fast rate, while the time to recuperate costs has grown shorter and shorter. These cost recuperation aspects have contributed to a large turn over in supercomputer manufacturing industry.

The thrust of the MPP architectures is "super number crunching." When efficient resource sharing, rather than super number crunching is the criteria, computer networks offer an advantage. With the availability of versatile workstations and desktop systems, computer networking has emerged as a popular architecture style.

This chapter summarizes some trends in computer systems architecture. Section 7.1 covers Dataflow architectures, a style that saw quite a bit of experimentation over the last two decades. It advocated the use of a large number of primitive processors that operate in parallel mode as invoked by the data requirements. No major commercial system of this type has emerged so far. Nevertheless, this style is a good example of architectures with tightly coupled massively parallel primitive processors. Section 7.2 introduces networks and the associated concepts of distributed processing, cluster computing, and grid computing, the latest trend in building massive information systems. Section 7.3 introduces architectures inspired by biology: neural networks based on brain models, DNA computing based on cellular biology and immunity-based systems. Optical technology has enhanced the performance of computer systems through fast data transmission rates of optical fibers and massive storage capabilities of optical disks. There have been numerous efforts to build optical computing devices with some recent successes in building optical switching devices. Section 7.4 provides a brief description of optical computing. These descriptions are necessarily very brief and the reader is referred to the books and journals listed at the end of this chapter for the latest on these and other architectures.

## 7.1 Dataflow Architectures

The basic structure underlying all the architectures described so far in this book is the one proposed by von Neumann. Parallel and pipeline structures were used to enhance the system throughput and overcome the limitations of the von Neumann model. Over the last few years, other types of architectures have evolved. The dataflow and reduction architectures introduced in Chapter 2 are two among many such experiments.

Recall that the von Neumann architectures are control *driven* and in them, data are passive and fetched only when an instruction needs them. That is, the operation of the machine is dictated by the instruction sequence (i.e., the program). Dataflow architectures on the other hand are *data driven*, since the readiness of data dictates the operation of the machine. That is, data are operated upon, when they are available. Reduction architectures are *demand driven* in the sense that data are fetched, when the evaluation of a function requires (i.e., demands) them. This chapter concentrates on dataflow architectures, since they have received much wider attention compared with other experimental architectures.

Dataflow architectures can be used advantageously for computation-oriented applications that exhibit a fine grain parallelism. Examples of such applications are image processing, automated cartography, and scene analysis. Studies conducted by National Aeronautics and Space Administration (NASA) have shown that this architecture type is suitable for aerodynamic simulation. Jack Dennis and other researchers at MIT have demonstrated that a dataflow computer can substantially increase the simulation speed over control flow computers. Consider for instance, the weather forecasting application in which a weather prediction model is used to predict the weather at some future time based on current and historical observations. A model implemented on a Control Data Corporation (CDC) 7600, took about 2 min to simulate 20 min of real-time weather. A dataflow model consisting of 256 processing elements, 128 addition units, 96 multiplication units, 32 memory modules, and a total memory of 3 MB, met the goal of 5 sec time for the same simulation.

Dataflow concepts have been utilized by various processors in their design. The CDC 6600 series of machines is an early example. Some Japanese manufacturers have introduced processor boards that utilize dataflow structures. Yet, after almost 20 years of experimentation, there are no major commercially available dataflow machines. Extensive hardware requirements and lack of progress in programming language and operating system development have been cited as the reasons.

### 7.1.1 Dataflow Models

As mentioned earlier, in a control flow architecture, the total control of the sequence of operations rests with the programmer. That is, a processor undergoes the instruction fetch, decode and execution phases for each instruction in the program. The data manipulation involves the transfer of data from one unit to another unit of the machine. Data are stored as variables (memory locations), taken to functional units, operated upon as specified by the programmer and the results are assigned to other variables.

A dataflow program on the other hand, is one in which the sequence of operations is not specified, but depends upon the need and the availability of data. There is no concept of passive data storage, instruction counter, variables, or addresses. The operation of the machine is entirely guided by data interdependencies and availability of resources. The architecture is based on the concept that an instruction is executable as soon as all its operands are available. As such, the data dependencies must be deducible from the program. Also, since the only requirement for operations to occur is the availability of operands, multiple operations can occur simultaneously. This concept allows the employment of parallelism at the operator (i.e., the finest grain) level. In fact, all dataflow architectures to-date have employed instruction-level parallelism.

---

**Example 7.1**
Figure 7.1 shows a Pascal program to calculate the roots of the quadratic equation $ax^2 + bx + c = 0$. In the conventional architecture, the execution follows the sequential structure of the program and on a single-processor machine, the following 8 steps are required:

```
1> a:=2 * a           ;must wait for input a
2> c:=b * b - 2 * a * c ;must wait for input b and step1
3> b:= -b/a           ;must wait for step1
4> c:=sqrt(c)         ;must wait for step2
5> c:=c/a             ;must wait for step4
6> a:=b + c           ;must wait for step5 and step3
7> b:=b - c           ;must wait for step5 and step3
8> output(a,b)        ;must wait for step6 and step7.
```

The number of steps can be reduced from 8 to 6 if multiple processors are available, since steps 2 and 3 and steps 6 and 7 can be performed simultaneously. The important characteristic of this program is that the sequence of operations is dictated by it.

```
Program find roots(input,output);
Var
    a,b,c : real;
begin
    writeln("Give the values of a,b,c");
    readln(a,b,c);
    a := 2 * a;
    c := b * b – 2 * a * c;
    c := sqrt(c);
    c := c/a;
    b := –b/a;
    a := b + c;
    b := b – c;
    writeln("The roots of the equation are");
    writeln(a,b);
end.
```

**FIGURE 7.1**
Pascal program to find the roots of the equation $ax^2 + bx + c = 0$

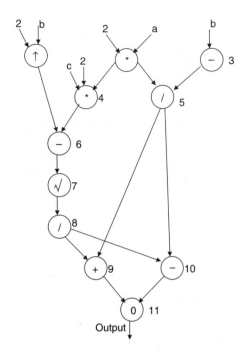

**FIGURE 7.2**
Dataflow graph for the solution of $ax^2 + bx + c = 0$

Now consider the *dataflow graph* (DFG) in Figure 7.2, representing the same computation. Here, the nodes (circles) represent operators (processors) and arcs represent paths that carry either data or control values between them. For example, Node 1 computes the square of its input operand "b" and its output becomes the input for Node 6. Node 3 negates "b" and its output is one of the inputs to Node 5. Node 2 multiplies its inputs "2" and "a" and its output "2 ∗ a" is placed on three arcs that become inputs to nodes 4, 5, and 8.

A node *fires* as soon as data are available on all its input arcs and its output arcs are free of data. The effect of node firing corresponds to removing the data from the input arcs, performing the node computation and placing the result on the output arc.

As soon as the values for a, b, c and the constant 2 are available, nodes 1, 2, and 3 fire simultaneously; nodes 4 and 5 fire as soon as they receive the output of node 2; node 6 fires after receiving data from nodes 1 and 4; and so on. Thus, arcs imply the data dependency of operations and each node starts computing as soon as data arrive on all its inputs. No other time or sequence constraints are imposed and several nodes can be active simultaneously. The important point to note is that there is no explicit representation of control flow in a DFG. The operations occur as and when data are ready.

Figure 7.3(a) shows a schematic view of a dataflow machine. The machine memory consists of a series of *cells*. A cell shown in Figure 3.7(b), provides all the information needed for an operation to occur (at an appropriate node

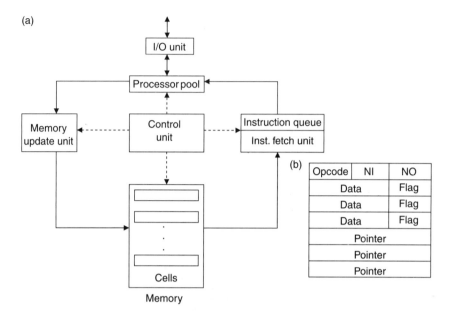

**FIGURE 7.3**
Static dataflow model: (a) structure and (b) cell

in the DFG). It consists of an operation code, number of input arcs (NI), number of output arcs (NO), NI input data slots, and NO output slots. Each data slot contains the data field and a flag that indicates whether the data value is ready or not. Each output slot points to a cell that receives the result of operations of that cell. When all the input operands are ready, a cell is firable. The Instruction Fetch Unit consists of multiple fetch units operating in parallel looking for firable cells in the memory. They fetch firable cells and convert them into instruction packets and enter them into the Instruction Queue. When a processor becomes available, it executes an instruction from this queue and presents the output to the Memory Update Unit, which in turn sends the results to appropriate memory cells based on the output pointers. This process continues until there are no firable cells. The I/O unit manages the interaction with external devices and the control unit coordinates the activities of all the units.

Now consider the details of the firing of a (multiplier) node shown in Figure 7.4. Here, the availability of the data on an arc is indicated by the presence of a *token*, on the arc (Rumbaugh, 1977). When a node has a data token on each of its input arcs, it fires and generates a data token on each of its output arcs. That is, when a node fires, the tokens on input arcs are removed and each output arc gets a token.

There are two node firing semantics commonly employed in a DFG: static and dynamic. In the *static* model, a node fires only when each of its input arcs has a token and its output arcs are empty. Thus, an acknowledgment of the fact that a subsequent node has consumed its input token is needed before

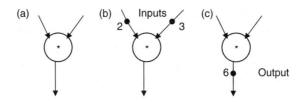

**FIGURE 7.4**
Node firing (a) no inputs, (b) all inputs are in, and (c) output produced.

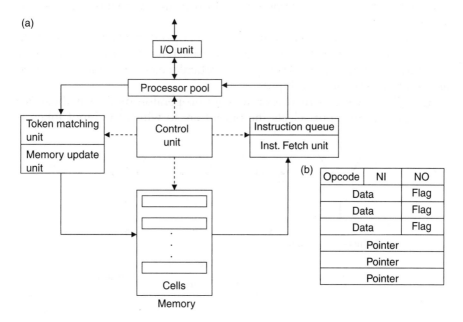

**FIGURE 7.5**
Dynamic dataflow model: (a) structure and (b) cell

a node in a static DFG can fire. The dataflow model of Figure 7.3 represents the static architecture. Here, each node receives as one of its input a control tag from the subsequent node indicating that the subsequent node has consumed the data on the output arc.

In the *dynamic* model, a node fires when all its inputs have tokens and the absence of token on its outputs is not necessary. Thus, it is possible to have multiple tokens on an arc. A tagging scheme is employed in dynamic DFGs to associate tokens with the appropriate dataset. The tag carries the information as to when and how a token was generated and which dataset it belongs to.

Figure 7.5 shows a dynamic dataflow machine. This is similar to the one in Figure 7.3 except for the token matching section. Here, the processors operate on the operands that have similar tags and the Memory Update Unit uses the tags to update cells belonging to the same dataset.

Note that a very high throughput can be achieved by dataflow architectures, since the algorithms are represented at the finest grain parallelism and hence the maximum degree of concurrency possible. This fine grain parallelism also poses a major challenge in the design and coordination of activities of various subsystems in the machine.

### 7.1.2 Dataflow Graphs

As seen by Figure 7.2, a DFG is a directed graph whose nodes represent operators and arcs represent pointers for forwarding the results of operations to subsequent nodes.

Figure 7.6 shows the three common types of operators in a DFG. The arithmetic and logical operators implement the usual binary operations. The array handling operator selects and outputs a particular element of the array based on the value of the selector input.

Figure 7.7 shows the structure of two control operators (nodes). In the *switch* node, the control input selects the output arc to which the input token is

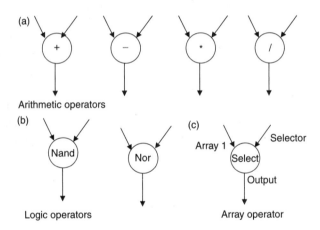

**FIGURE 7.6**
DFG operators

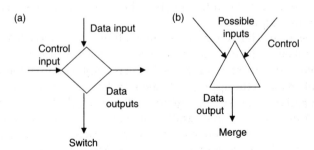

**FIGURE 7.7**
DFG control operators

transmitted. In the *merge* node, the control input selects one among the data inputs to be placed on the output. In these cases, it is not necessary that the inputs not selected have a token, before the nodes fire.

### Example 7.2
Figure 7.8 shows a loop construct implemented as a DFG. The "Condition" operator evaluates the loop control condition and generates the control input for the switch node which either allows the continuation of the loop or exits from it.

The classical dataflow model is represented by the *cyclic* graphs of the type shown in Figure 7.8 (while the functional model is represented by the *acyclic* graph notation). A major problem of cyclic graphs is the "race" condition, as illustrated by the following example.

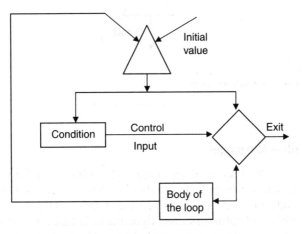

**FIGURE 7.8**
Cyclic graph for loop control

### Example 7.3
In the graph shown in Figure 7.9, two computations produce the values A and B. These values are used by a node which also receives the values X and Y produced by another set of two nodes. Ideally, A needs to be combined with X and B with Y. If the node producing A as the output takes longer than the one that produces B, then B meets with X and A meets with Y, which is not desirable. To eliminate this problem, labels are attached to the data in order that only those which belong to a certain

label can pair together. A checking procedure is then needed in the computations to ensure that both the operands carry the same label.

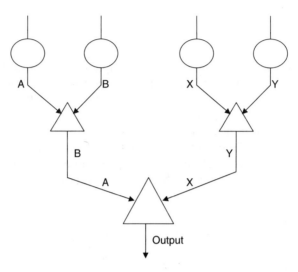

**FIGURE 7.9**
Race condition in a dataflow graph

### 7.1.3 Dataflow Languages

The primary goal of a high-level language (HLL) for the dataflow computers is the representation of parallelism and the dataflow as depicted by a DFG. The best known dataflow languages are: Irvine dataflow language (Id) (Arvind and Nikhil, 1989) and value-oriented algorithmic language (VAL) (Ackerman, 1982). Other dataflow languages are: HASAL (Sharp, 1985), Lapse (Glauert, 1978) and SISAL (streams and iterations in single assignment language) (Gaudiot and Bic, 1989). There have been attempts to use conventional languages to program dataflow machines (as in a Texas Instruments' design that used Fortran). On the other hand, the goal of languages like SISAL was to define a "Universal" language for programming multiprocessor systems.

The essential features of a dataflow language are

1. *The language should be functional*: The term "functional" implies a language that operates by applications of functions to values. In a functional language, all control structures are replaced by combining operators that manipulate functions directly, without ever appearing to explicitly manipulate data. A functional programming (FP) system consists of a set of objects, functions, functional forms and definitions, and a single operation namely application. The power of an FP system

is due to its ability to combine functions to form new functions. An example of a functional language is (pure) LISP.

2. *The language should allow a nonsequential specification*: That is, the order of the function definitions should not be related to their order of execution, which is a fundamental property of the dataflow system.

3. *The language should obey the single assignment rule*: That is, the same variable name cannot be used more than once on the left-hand side of any statement. A new name will be given for any redefined variable and all subsequent references shall be changed to the new name, as illustrated in the following example.

---

**Example 7.4**
The sequence of statements:

$X := A + B$
$Y := A - B$
$X := X * Y$

has to be changed to the following:

$X := A + B$
$Y := A - B$
$Z := X * Y$

---

Single assignment rule facilitates the detection of parallelism in a program and offers clarity, compared with repeated use of the same name. Tesler and Enea (1968) proposed this rule and it was applied in the development of the French dataflow computer LAU and the Manchester dataflow machine from England. (The optimizing compilers on conventional machines also perform this transformation.)

4. There should be no side effects, since a node in a dataflow graph is self-contained and there is no global data space of values or use of shared-memory cells.

Dataflow languages differ from conventional languages in several aspects:

1. The concept of variables is much different in dataflow languages compared with that in a control flow language. They are not assigned values in the sense that a value is routed to a specific memory location reserved for that variable. That is, all variables are values (including arrays) and not memory locations. Operations merely

produce new values. Names are assigned to values, instead of values being assigned to names. The names are for the convenience of the programmer. Since the values produced by a computation are to be used immediately after they are produced, they can be simply sent to the next computation for which it is intended.

2. Dataflow languages are "applicative" in nature. That is, values are operated on to produce more values. These values in turn are used for other operations until the given task is completed.

3. One property of dataflow languages is the "locality" of effect. When data is produced by one node of the dataflow graph it is used by some other node, but does not influence the computation performed another node. Because of this property, temporary names can be used in mutually exclusive section of a program. Different sections may use the same name and still execute concurrently without influencing the computations performed by other sections. This is true of any block structured language.

4. Because of the data-driven mode of execution, GO TO constructs of conventional languages are not required in dataflow languages. The DFG contains all the information necessary to execute the program.

5. The iteration structures of dataflow languages are somewhat unusual because of "no side effects" and "single assignment" properties. These languages are not particularly suited to execute loops and usually only one type of iteration structure is allowed.

Dataflow languages can be classified as either *Graphical* or *Textual*. The DFG notation used so far in this chapter represents graphical languages, while VAL and Id are examples of textual languages.

There are two graphical notations: *cyclic* and *acyclic*. The classical model of dataflow is represented by the cyclic graph notation, while the functional model is represented by the acyclic graph notation.

As noted in earlier chapters, efficient manipulation of data structures is a difficult task for any parallel environment. It poses special problems in dataflow machines because of the use of functional languages (with no side effects and no updatable storage), data-driven computation and a very high degree of parallelism.

All dataflow machines have utilized the concept of a structure store, which holds the values of the elements of the data structure, while labels carry their addresses. Different machines allow different types of operations on these structure stores.

### 7.1.4 Example Dataflow Systems

This section provides a brief description of two experimental systems implementing the static and dynamic dataflow models. These systems have been implemented in one of the two forms: Ring and Network.

In a *Ring* architecture, the processors are connected together by a ring. All the processors execute tasks independently and produce results. The results are sent to the control unit which assigns them to subsequent operations, and generates new tasks which are passed around the ring. The architecture may contain either a single ring or multiple rings connected together in parallel layers. In the latter case, a switch is provided to enable communication between the rings. Also, the control function may be distributed between different layers.

In a *Network* architecture, a separate control unit is not used. The processing nodes themselves perform the control task, and are interconnected by a network.

### 7.1.4.1   MIT static architecture

In the static dataflow model, only one data token is allowed to exist on an arc at any given time. Also, control tokens are needed for proper transfer of data tokens from node to node. The MIT static architecture, the Data-Driven Processor (DDP) of Texas Instruments (TI) Incorporated and the French LAU system are the popular static architectures.

A schematic view of the MIT static architecture is shown in Figure 7.10. It consists of five major blocks connected by channels through which information flows in the form of packets. The memory contains the cells representing the nodes of a DFG. It receives the data tokens from the distribution network

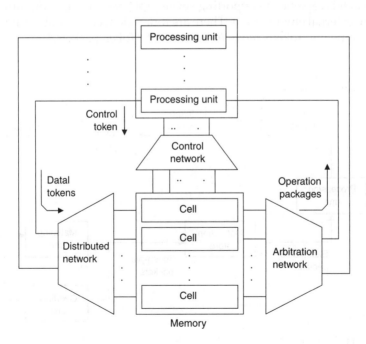

**FIGURE 7.10**
MIT static dataflow machine

and control tokens from the control network. When all the operands in a cell have arrived, the cell becomes an operation packet (i.e., enabled instruction). The arbitration network transfers the operation packets to the processing section. The processing units in the processing section perform functional operations on data tokens. The results of the execution are transferred to the memory through the distribution and the control networks. They become the operands for subsequent instructions.

### 7.1.4.2 Manchester dataflow machine

The main characteristic of the dynamic architectures is that, many data tokens can exist on the same arc at any instant of time. These data tokens need to be tagged, so that only those tokens which belong to the same pair can be operated on. These tags can be either labels or colors and there is no need for control tokens acknowledging the transfer of data tokens between the nodes. Tagging of data tokens and their matching requires extra hardware. Irvine dataflow machine developed by Arvind (Arvind and Nikhil, 1989), the Manchester dataflow machine (MDM) developed by Watson and Gurd at Manchester, England, the Data-Driven Machine (DDM) proposed by Davis (1978), the Epsilon dataflow processor of the Sandia National Laboratories, the Experimental system for Data Driven processor arraY (EDDY) of Japan and the MIT Monsoon system are examples of dynamic architectures.

The MDM shown in Figure 7.11, is organized as a circular pipeline. The I/O switch is capable of supporting seven pipelines and also communicates with the external environment. There are four independent units forming the circular pipeline. Tokens move in the form of "Token packets."

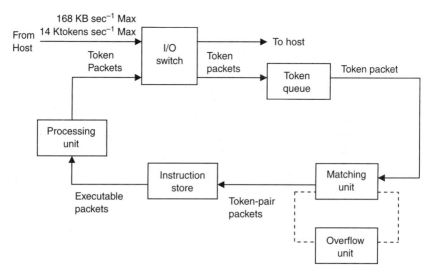

**FIGURE 7.11**
Manchester dataflow machine

The token packets first enter the Token Queue, which is used to smooth out the uneven rates of the generation and consumption of tokens. It consists of three pipeline buffer registers surrounding a first-in-first-out (FIFO) circular store with a capacity of 32K tokens and with an access time of 120 nsec. The width of the store and the registers is 96 bits and the clock period is 37.5 nsec. The maximum throughput of the queue is 2.67 million tokens sec$^{-1}$.

The tagged token matching is done in the matching unit which matches the tokens with the same tags (i.e., the token pairs designated for the same instruction). The single operand instructions bypass this unit. The matching utilizes a hardware hashing mechanism and consists of an associative memory capable of holding up to 1.25 million unmatched tokens awaiting their partners. This unit has 8 buffers and 20 64K-token memory boards, which also contain a comparator. The incoming token has a 54-bit matching key, where 18 bits indicate the destination and 36 bits are for the tag. This will be hashed to a 16-bit address corresponding to the same cell on each board. Each board compares the key of the token stored at that address against the matching key of the incoming token. If a match is found, the two tokens are combined into a 133-bit Group package and will be sent to the Instruction store; if no match is found, the incoming token is written at that address on one of the 20 boards. Thus, the system can concurrently accommodate 20 tokens that hash to the same address. The subsequent tokens are sent to the Overflow unit to continue their search.

The Instruction store is a buffered random access memory with a capacity of 64K instructions. The memory access time is 150 nsec, and the clock period is 40 nsec, providing a maximum rate of 2 million instruction fetches sec$^{-1}$.

The instructions are fetched from the store and executed by the processing unit resulting in the production of new tokens, which may reenter the queue. The token package entering and leaving the processing element is 96-bits wide and consists of 37 bits of data, 36 bits for the tag, and 23 bits for destination address, input port number and the number of tokens that are needed to enable the destination instruction.

The processing unit consists of a homogeneous bank of 15 microcoded functional units. Each functional unit has 51 internal registers and 4K words of writable microcoded memory. The internal word length is 24 bits.

The dataflow graphs that are executable on this machine are generated by a compiler from the HLL "Lapse," a Pascal-like language.

The prototype of this machine had a single processing element and two structure stores (structure rings) connected to a host (VAX 11/780) with an I/O switching network. The speed of the link connecting the host was 168 KB sec$^{-1}$ (14K tokens sec$^{-1}$). The two structure store modules hold a total of 1 million data values with an access rate of 0.75 million reads sec$^{-1}$ and nearly half as many writes sec$^{-1}$. Arrays and other data structures were stored in this memory as "I-Structures," which improve the parallelism and efficiency of computations involving data structures, by allowing these structures to be created incrementally. This enables the production and consumption of a data structure to proceed concurrently.

### 7.1.4.3 MIT/Motorola monsoon system

The Monsoon Dataflow Multiprocessor, built jointly by MIT and Motorola, is a small shared-memory multiprocessor. The motivation behind building this research prototype was to demonstrate the feasibility of general-purpose parallel computers, suitable for both numerical and symbolic applications. The main goal of this project was to show that one could easily write parallel programs in an implicitly parallel programming language (Id) and that these programs would run efficiently. Several programs written in Id previously have been implemented on the Monsoon system with eight processors and have achieved speedups of more than seven. Since the Monsoon processor consists of an eight-stage interleaved pipeline, an 8-processor Monsoon requires at least 64-fold parallelism to achieve 100% utilization.

As detailed earlier, there are three levels of parallelism: processor level — consisting of procedure activations on separate processors, thread-level — consisting of multiple threads active on each processor, and instruction-level — consisting of instructions within a thread. The Monsoon system, being a dataflow architecture, tries to exploit all three types of parallelism. Refer to Hicks et al. (1992) for further details.

Dataflow research has received a wide attention. The most notable efforts are at MIT (Jack Dennis and his research associates on static architectures and Arvind and his group on dynamic architectures); at the University of Manchester, England by John Gurd and Ian Watson which lead to the development of the Manchester dataflow machine, one of the earlier dataflow architectures; at Sandia National Laboratories, New Mexico; University of Utah; France and Japan. The Ministry of International Trade and Industry of Japan has sponsored dataflow research, and Hitachi Ltd. is privately funding research at the University of California at Berkeley.

The major reasons for this architecture not being a commercial success are: extensive hardware complexity, complexity of representation schemes, and difficulty in developing and compiling languages and other representation schemes. There is much controversy regarding the viability of this architecture type. Some feel that this type of architecture is just a laboratory curiosity and can never be implemented in a cost-effective manner. The others argue that with the availability of faster VLSI components and the advances in dataflow representation techniques, it is possible to build cost-effective dataflow machines.

Japanese manufacturers seem to be leading the field in dataflow hardware technology. One of the earliest machines is the SIGMA-1 from the Electrotechnical laboratory (ETL) in Tsukuba. The development of the EM series of machines began in 1981 at ETL. The current version EM-5, employs a hybrid architecture composed of dataflow and control flow concepts. It is expected that a 1000 processor machine would be built by the end of 1993. The aim of the EM project is to develop a general-purpose machine with 16K, 80 MIPS processors. In 1991, Japan's Sharp corporation announced a data-driven processor (DDP), produced jointly with Mitsubishi. This is a 1024-node architecture and uses a C-like programming language. The Enhanced Data-Driven Engine (EDDEN) of Sanyo electric company is also a 1024-node architecture, interconnected by a two-dimensional torus network. The nodes are based on a 32-bit custom-designed processor. The commercial version of this architecture (Cyberflow) announced by Sanyo in 1992 is a desk top system with 4 to 64 nodes and a peak rating of 640 MFLOPS.

## 7.2   GRID Computing

Pinfold (1991) identified three areas in which conventional SIMD and MIMD approaches fall short:

1. They are not highly scalable and hence cannot be easily expanded in small increments.
2. They employ only one type of general-purpose processor and hence are not suitable for environments with an array of specialized applications.
3. They employ fixed interconnection topology, thereby restricting the users when the applications dictate a different more efficient topology.

The SIMD and MIMD systems built since 1991 have addressed the first two shortfalls by using heterogeneous processing nodes and being scalable to fairly large number of nodes. They have also merged the SIMD and MIMD concepts, as evidenced by the evolution of the Thinking Machines CM series. The earlier machines in the series were SIMDs while the CM-5 operates in both the modes.

*Computer networks* are the most common multiprocessor architectures today. Figure 7.12 shows the structure of a computer network. It is essentially a message-passing MIMD system, except that the nodes are loosely coupled, by the communication network. Each node (Host) is an independent computer system. The user can access the resources at the other nodes through the network. The important concept is that the user executes his application at the node he is connected to, as far as possible. He submits his job to other nodes when resources to execute the job are not available at his node. The Internet is the best example of the worldwide network.

A *distributed processing system* utilizes the hardware structure of the network shown in Figure 7.12. It is more general in its operation in the sense that the nodes operate in what is called a "cooperative autonomy." That is, the nodes cooperate to complete the application at hand and all the nodes are equal in capability (i.e., there is no master/slave relation among the nodes). In a distributed system, the hardware, the data, and the control (operating system) are all distributed.

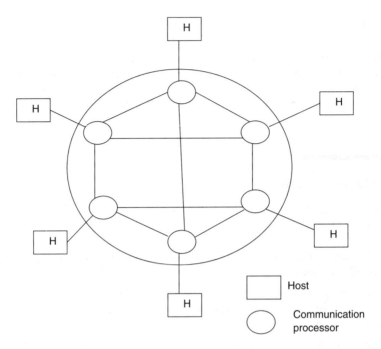

**FIGURE 7.12**
Computer network

The next step in this progression of building powerful computing environments is *Grid computing*. A grid is a network of computers that act as a single "virtual" computer system. Utilizing specialized scheduling software, grids identify resources and allocate them to tasks for processing on the fly. Resource requests are processed wherever it is most convenient, or wherever a particular function resides, and there is no centralized control. Grid computing exploits the underlying technologies of distributed computing, job-scheduling, and workload management which have been around more than a decade. The recent trends in hardware (commodity servers, blade servers, storage networks, high-speed networks, etc.) and software (Linux, Web services, open source technologies, etc.) have contributed to make grid computing practical. Hewlett–Packard's Adaptive Enterprise initiative, IBM's On Demand Computing effort, Sun Microsystems's Network One framework are examples of commercial grid computing products.

There are several definitions of "grid computing architecture." Some define clustered servers that share a common data pool as grids. Others define large distributed networked environments that make use of thousands of heterogeneous information systems and storage subsystems as grids. There are definitions that fall somewhere between these two ends of the spectrum. A grid in general is a network architecture for connecting compute and storage resources. It is based on standards that allow heterogeneous systems and applications to share compute and storage resources transparently.

A computational grid is a hardware and software infrastructure that provides dependable, consistent, pervasive, and inexpensive access to high-end computational capabilities (Foster and Kesselman, 2004). The resource sharing is not primarily file exchange but rather direct access to computers, software, data, and other resources, as required by a range of collaborative problem-solving and resource-brokering strategies emerging in industry, science, and engineering. This sharing is necessarily, highly controlled, with resource providers and consumers defining clearly and carefully just what is shared, who is allowed to share, and the conditions under which the sharing occurs (Foster et al., 2001). Thus, a grid is a system that coordinates resources that are not subject to centralized control, uses standard, open, general-purpose protocols and interfaces, and delivers nontrivial qualities of service.

A grid consists of networks of computers, storage, and other devices that can pool and share resources and the grid architecture *provisions* those resources to users and applications. All the elements in the grid resource pool are considered to be *virtual* (i.e., they can be activated when needed).

*Provisioning* involves locating computing and storage resources and making them available to requestors (users/applications). Some ways to provision resources are: reconfiguring to meet the new workload; repartitioning an existing server environment to provide additional storage; activating installed but not active components (CPU, memory, or storage) on a local server to provide local access to additional resources; accessing additional computing and storage resources that can be found and exploited over a local area network or across a wide area network; and finding and exploiting additional computing and storage resources using a grid.

All computing and storage resources are considered virtual and become *real* when they are activated. Two approaches to exploiting virtualized resources on a local server environment are: manual or programmatic reconfiguration, and activation of compute/storage/memory modules that have been shipped with systems (i.e., activated and paid for when used). Virtualized resources can also be found on a network. For instance, clustered systems provide access to additional CPU power when needed (utilizing a tightly coupled network) and, grid software can find and exploit loosely coupled virtual resources on a local or a wide area network. Also, virtualized services can be acquired externally using a utility computing model.

Grids are generally deployed at the department level (departmental grid), across an enterprise (intergrids) and, across multiple enterprises or between multiple organizations (extragrids). Sun Microsystem's scalable virtual computing concept identifies three grid levels:

- Cluster Grid (Departmental computing): simplest grid deployment, maximum utilization of departmental resources, resources allocated based on priorities.
- Enterprise Grid (Enterprise computing): resources shared within the enterprise, policies ensure computing on demand, gives multiple groups seamless access to enterprise resources.

- Global Grid (Internet computing): resources shared over the internet, global view of distributed datasets, growth path for enterprise grids.

In 2001 the Global Grid Forum (the GGF — the primary grid standards organization) put forward an architectural view of how grids and Web services could be joined. This architecture is called the Open Grid Services Architecture or OGSA. Since the GGF announced its OGSA view there has been strong progress in the articulation of Web services and grid standards. The following organizations are the leading standards organizations involved in articulating and implementing Web services and service oriented architectures (SOA) over grid architecture:

- The GGF is the primary standards setting organization for grid computing. The GGF works closely with OASIS (described below) as well as with the Distributed Management Task Force to help build interoperable Web services and management infrastructure for grid environments.
- The Organization for the Advancement of Structured Information Standards (OASIS) is very active in setting standards for Web services, and works closely with the GGF to integrate Web services standards with grid standards.
- The Distributed Management Task Force (DMTF) works with the GGF to help implement DMTF management standards such as the DMTF's Common Information Model (CIM) and Web-Based Enterprise Management (WBEM) standards on grid architecture.
- The World Wide Web Consortium (W3C) is also active in setting Web services standards (and standards that relate to XML).
- The Globus Alliance (formerly The Globus Project) is also instrumental in grid standards — but from an implementation point-of-view. The Globus Alliance is a multi-institutional grid research and development organization. It develops and implements basic grid technologies and builds a toolkit to help other organizations implement grids, grid standards, and even OGSA proposed standards.

### 7.2.1 Open Grid Services Architecture

The OGSA is an architectural vision of how Web services and grid architectures can be combined. There are four working groups that help implement OGSA standards. These groups focus on defining clear programmatic interfaces, management interfaces, naming conventions, directories, and more. The specific OGSA working groups involved in these activities are: Open Grid Services Architecture Working Group (OGSA-WG), Open Grid Services Infrastructure Working Group (OGSI-WG), Open Grid

Service Architecture Security Working Group (OGSA-SEC-WG) and, Database Access and Integration Services Working Group (DAIS-WG). The OGSI is an implementation/testbed of OGSA.

There are several standards involved in building an SOA and underlying grid architecture that can support business process management. These standards form the basic architectural building blocks that allow applications and databases to execute service requests. And these standards also make it possible to deploy business process management software that enables Information Systems (IS) executives to manage business process flow. The most important grid and grid-related standards include:

- Program-to-program communications (SOAP, WSDL, and UDDI)
- Data sharing (extensible markup language — XML)
- Messaging (SOAP, WS-Addressing, MTOM (for Attachments))
- Reliable Messaging (WS-Reliable Messaging)
- Managing workload (WS-Management)
- Transaction-handling (WS-Coordination, WS-Atomic Transaction, WS-Business-Activity)
- Managing resources (WS-RF or Web services resource framework)
- Establishing Security (WS-Security, WS-Secure Conversation, WS-Trust, WS-Federation, Web Services Security Kerberos Binding)
- Handling metadata (WSDL, UDDI, WS-Policy)
- Building and integrating Web services architecture over a grid (see OGSA)
- Orchestration (standards used to abstract business processes from application logic and data sources and set up the rules that allow business processes to interact)
- Overlaying business process flow (business process engineering language for Web services — BPEL4WS)
- Triggering process flow events (WS-Notification)

Grids are being used in a variety of scientific and commercial applications such as: aerospace and automotive (for collaborative design and modeling), architecture (engineering and construction), electronics (design and testing), finance (stock/portfolio analysis, risk management), life sciences (data mining, pharmaceuticals), manufacturing (inter/intra-team collaborative design, process management), and media/entertainment (digital animation). Some of the most famous scientific and research grids include:

- The Seti@Home Project — thousands of Internet PCs used for the search for extraterrestrial life).
- The Mersenne Project — the Great Internet Mersenne Prime Search (GIMPS) is a worldwide mathematics research project.

- The NASA Information Power Grid — the IPG joins supercomputers and storage devices owned by participating organizations into a single, seamless computing environment. This project will allow the government, researchers, and industry to amass computing power and facilitate information exchange among NASA scientists).
- The Oxford e-Science Grid — Oxford University's "e-Science" project addresses scientific distributed global collaborations that require access to very large data collections, very large scale computing resources, and high-performance visualization back to the individual user scientists.
- The Intel-United Devices Cancer Research Project — this project is a grid-based research project designed to uncover new cancer drugs through the use of organizations and individuals willing to donate excess PC processing power. This excess power is applied to the grid infrastructure and used to operate specialized software. The research focuses on proteins that have been determined to be a possible target for cancer therapy.

The largest grid effort currently underway is the "TeraGrid" scientific research project. The TeraGrid was launched by the United State's National Science Foundation in August 2001 as a multi-year effort to build the world's largest grid infrastructure for scientific computing. In 2004, the TeraGrid will include 20 teraflops of computing power, almost one petabyte of data, and high-resolution visualization environments for modeling and simulation. The supporting grid network is expected to operate at 40 GB $\text{sec}^{-1}$.

Although the preponderance of compute grids have been in the scientific, research, and educational communities, there is a strong growth of compute grids in commercial environments.

At the end of 2003, the Office of Science of the U.S. Department of Energy published a report called "Facilities for the Future of Science: A 20-Year Outlook" located at http://www.er.doe.gov/Sub/Facilities_for_future/20-Year-Outlook-screen.pdf. This report details how the U.S. government will use ultrascale computing (a very high-speed grid approach with very powerful servers/supercomputers) to encourage discovery in the public sector.

India has undertaken the building of a national "I-Grid" (information grid). India's Centre for Development of Advanced Computing — makers of India's PARAM Padma supercomputers — sees its role as helping India to carve out a niche in the global arena of advanced information technology, to further expand the frontiers of high-performance computing, and to utilize resulting intellectual property for the benefit of society "by converting it into an exciting business opportunity and establishing a self-sustaining and wealth creating operation."

The United Kingdom has created E-Science Centers that utilize data grids for scientific research projects. The National E-Science Center coordinates projects with regional centers located Belfast, Cambridge, Cardiff, London, Manchester, Newcastle, Oxford, and Southampton, as well as with sites in Daresbury and Rutherford Appleton. These centers provide facilities to scientists and researchers who wish to collaborate on very large, data-intensive projects. This project is one of dozens of governmental grid projects within the United Kingdom.

Numerous other examples of government grids can be found at http://www.grids-center.org/news/news_deployment.asp.

## 7.3 Biology Inspired Computing

Three aspects of human biology have inspired building of computer systems. Researchers in cognitive science have developed models of brain's computational activity in the form of neural networks. The cell behavior has inspired a field of computing based on DNA (Deoxyribose Nucleic Acid) structures and models. An examination of the body's immune system behavior has resulted in the concept of artificial immune systems. The neural network concepts have already resulted in sophisticated architectures, while the other two concepts are now active areas of research interest. This section provides a brief introduction to these three computing models.

### 7.3.1 Neural Networks

Neural networks or more specifically artificial neural networks (ANN) are suitable for solving problems that do not have a well defined algorithm to transform an input to an output. Rather, a collection of representative examples of the desired transformation is used to "train" the ANN, which in turn "adapts" itself to produce the desired outputs, when presented with the example inputs. In addition, it will respond with an (good guess) output even when presented with inputs that it has never seen before. That is, during the training mode, the information about the problem to be solved is encoded into the ANN and the network spends its productive mode in performing transformations and "learning." Thus the advantage of the ANN approach is not necessarily in the elegance of the particular solution, but in its generality to find a solution to particular problems, given only examples of the desired behavior. It allows the evolution of automated systems without explicit reprogramming. This section provides a brief description of the state-of-the-art in neural network technology.

The ANNs resemble their biological counterparts in terms of their behaviors of learning, recognizing, and applying relationships between objects. Biological neural networks consist of a large number of computational elements called neurons. A *neuron* is composed of the cell body, a number of extensions called *dendrites* which serve as inputs and a long extension called *axon* which serves as the output. *Synapses* connect the axon of one neuron to dendrites of other neurons. The neurons are arranged in layers. In general, neurons in one layer receive their inputs from those in another layer and send their outputs to the neurons in a third layer. Depending on the application, it is possible that the neurons in a layer receive inputs from and provide outputs to the neurons in the same layer. Neurons undergo a constantly changing state of biochemical activity. The composite state of all the neurons in a layer constitutes the representation of the world. The connections between the neurons have weights associated with them. These weights (called coupling weights) represent the influence of one neuron over the other connected to it. If two neurons are not connected the corresponding coupling weight is 0. Each neuron essentially sends its state information (i.e., value) multiplied by the corresponding coupling weight to all the neurons it is connected to. All neurons sum the values received from their dendrites to update their state.

The ANNs are thus based on a threshold processing circuit or neuron (shown in Figure 7.13) at which the weighted inputs are summed. Here, the weighting functions have been relabeled as attenuators. In practical implementations, the inputs to a neuron are weighted by multiplying the input by a factor which is less than or equal to one, that is, attenuated. The value of the weighting factors are determined by the learning algorithm. The attenuated inputs are summed using a nonlinear function called a sigmoid function. If the output of the summing function exceeds a built in threshold, the neuron "fires," generating an output.

Figure 7.14 shows the layered model of an ANN consisting of a large number of neurons. Each neuron has multiple inputs and its output is connected to a large number of other processor's inputs. As mentioned earlier, in the normal operating mode of the ANN, data presented as input will cause a specific output pattern to be generated. The input to output relationship is determined during the "training mode," when a known input is presented

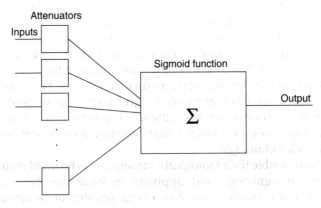

**FIGURE 7.13**
Neuron model

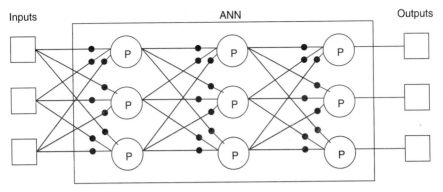

**FIGURE 7.14**
ANN structure

along with the expected output. The training algorithm adjusts the weighting of inputs until the expected output is attained.

Some of the most common ANN architectures are described below.

### 7.3.1.1 Multilayered perceptron networks

Multilayer perceptrons (MLPs) are feedforward neural networks trained with the standard back propagation algorithm. They comprise a set of source nodes that forms an input layer, one or more hidden layers of computation nodes and an output layer of nodes. The input signal propagates through the network layer-by-layer. A typical three layer MLP network is shown in Figure 7.15.

This type of architecture is extensively used in pattern classification, where the network is made to learn to transform input data into a desired response. "Back propagation" algorithm is typically used to train the MLP neural networks. This algorithm consists of two steps. In the *forward pass*, the predicted outputs corresponding to given inputs are evaluated using some standard equation. In the *backward pass*, partial derivatives of the cost function with respect to different parameters are propagated back through the network. The chain rule of differentiation gives similar computational rules for the backward pass as the ones in the forward pass. The network weights are then adapted using any gradient-based optimization algorithm. The whole process is iterated until the weights have converged.

### 7.3.1.2 Radial basis function networks

The radial basis function (RBF) networks were introduced by Broomhead/Lowe and Poggio/Girosi in the late 1980s. They are motivated by the locally tuned response observed in biologic neurons in the visual or auditory system. RBFs have been studied in multivariate approximation theory, particularly in the field of function interpolation. The RBF neural network model

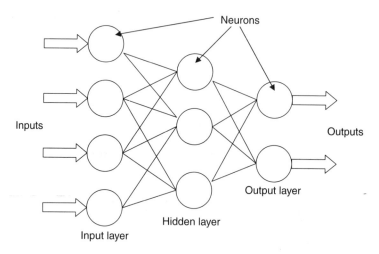

**FIGURE 7.15**
Three layer fully connected MLP network

is an alternative to MLP which is perhaps the most often used neural network architecture.

The RBF network is similar to MLP as it is a multilayer, feedforward network. However, hidden units in RBF are different from the one in the MLP. They contain "Radial Basis Function" which is a statistical transformation based on Gaussian distribution. Each hidden layer takes as its input all the outputs of the input layer $x_i$. The hidden unit has a "basis function" which has the parameters "center" and "width." The center of the basis function is a vector of numbers $c_i$ of the same size as the inputs to the unit and there is normally a different center for each unit in the neural network. Unit computes the "radial distance," $d$, between the input vector $x_i$ and center $C_i$ using the Euclidean distance:

$$d = \text{SQRT}((x_1 - c_1)^2 + (x_2 - c_2)^2 + \cdots + (x_n - c_n)^2)$$

The unit output $a$ is then computed by applying the basis function $B$ to this distance divided by the width $w$:

$$a = B(d/w)$$

Thus the basis function is a curve that has a peak at zero. The structure of the hidden layer unit in an RBF network is shown in Figure 7.16.

One of the basic differences between RBF and MLP is that RBF neural network is easier to train. However, the disadvantage of RBF networks is its greater computational complexity and, often, more manual tuning before learning is robust and efficient. Moreover it is not very well suited for the large applications.

# Current Directions

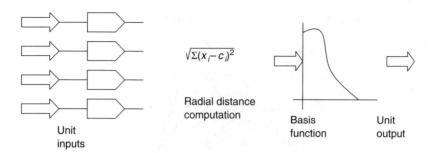

**FIGURE 7.16**
Hidden layer of RBF network

### 7.3.1.3 Recurrent neural networks

Recurrent neural networks (RNNs) is a network of neurons with feedback connections. They are different from the feed-forward networks where the outputs from the output layer are fed back to the input units. The main advantage of such networks is that they can store information about time and thus are being used for predicting different types of time series. RNN contain five layers: input and output layers and three hidden layers. All the information regarding the previous inputs to the neural network and the predictions made are stored in the hidden layers. As shown in Figure 7.17, one layer holds the last input, another the vector corresponding to the previous output state and the third forms the output vector, corresponding to the current prediction.

One of the applications of the RNN is in continuous Speech recognition. For details see Robinson et al. (1995).

### 7.3.1.4 Auto-associative neural networks

Auto-associative neural networks are the ones where the two MLPs are connected "back to back" as shown in Figure 7.18.

Auto-associative neural networks are trained with the target dataset that is identical to the input dataset. The number of units in the middle layer must be less than the number of inputs (and outputs). In training, the network weights are adjusted until the outputs match the inputs, and the values assigned to the weights reflect the relationships between the various input data elements. The disadvantage of conventional auto associative neural networks is that they are inefficient. The effect of training is to adjust the weights to values that are best for most patterns. At the end of training, all weights are fixed to reflect the majority of patterns. All the patterns that represent minorities (from the perspective of a single weight) are ignored. These networks are mainly used in data compression applications.

### 7.3.1.5 Self-organizing maps

Self-organizing maps (SOMs) are a data visualization technique invented by Teuvo Kohonen. Since humans cannot visualize high-dimensional data, they

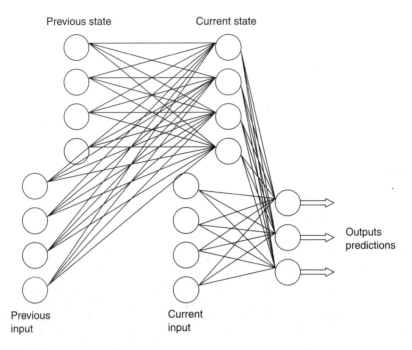

**FIGURE 7.17**
Recurrent neural networks

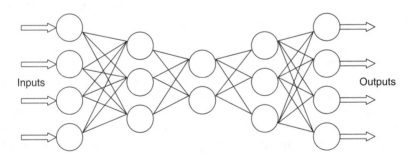

**FIGURE 7.18**
Auto-associative neural networks

reduce the dimensions through the use of a self-organizing neural network. It reduces dimensions by producing a map of usually 1 or 2 dimensions that plot the similarities of the data by grouping similar data items together. Thus it reduces dimensions along with displaying similarities and hence is useful in applications where it is important to analyze a large number of examples and identify groups with similar features. The learning process is competitive and unsupervised.

The SOM is widely used as a datamining and visualization method for complex datasets. Application areas include image processing and speech

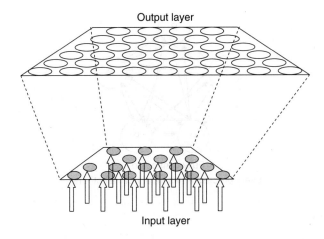

**FIGURE 7.19**
A Kohonen network

recognition, process control, economical analysis, and diagnostics in industry and in medicine.

Figure 7.19 shows a Kohonen network. The feature output map represents a continuous surface where the edges wrap round to join up at the opposite side. Thus it can be viewed as a topologically organized toroidal surface. These features need not be restricted to two dimensions, and a Kohonen network can be designed to create a feature map with more than two dimensions, if required by the particular problem.

### 7.3.1.6 Implementation technologies

Note that the neurons in Figure 7.14 have a very modest computational complexity and they normally communicate only with the nearest neighbors making the interconnections simple and repetitive. Because of these characteristics and the capabilities offered by the VLSI technology it is now possible to build ANNs with a large number of processors cost effectively.

There are two types of connectivity models encountered in the practical implementation of ANNs. The first, the Hopfield model illustrated in Figure 7.20, is a fully connected model in which the output of every neuron is an input to every other neuron. This is the most difficult model to implement because of the exponential increase in network complexity as the number of neurons grows. The second is the layered model illustrated in Figure 7.14. This model requires fewer attenuators and interconnections and enables multichip implementations. The number of layers and the number of neurons found in each layer varies from design to design.

Several technologies have been used in implementing ANNs. These include digital electronics, analog electronics, analog/digital (hybrid) designs, optical technology and simulation. The main reasons for digital implementation are

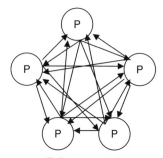

Fully connected
Every processor has input from all others

**FIGURE 7.20**
Hopfield model

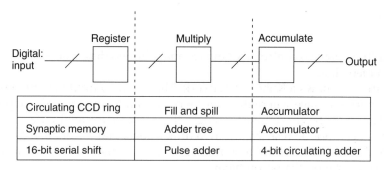

**FIGURE 7.21**
Digital implementations

the ability to readily program weights, high noise immunity, and flexibility for signals to be transmitted between chips. But, digital implementations occupy a large silicon area because of the synaptic multipliers. Figure 7.21 illustrates the basic elements which comprise a digital ANN and a few implementation techniques. Parallel input data (usually 8–16 bits) is clocked into a register, which is then multiplied with the synaptic weight (stored digitally) and accumulated using some form of digital adder mechanism. The digital storage of synapse weights are stable over a long period and do not suffer from temperature variations. It is relatively easy to load or retrieve trained configuration data to/from the circuitry.

A disadvantage of digital implementations is that due to the more extensive circuitry required, the number of neurons per package is much less compared to analog designs. Digital ANNs must be clocked through their multiply and accumulation cycles, which greatly reduces the throughput as compared to analog implementations. Another problem is that inputs to digital ANNs must be presented in the form of digital data. This requires converting analog data sources to a digital form compatible with the ANN design.

# Current Directions

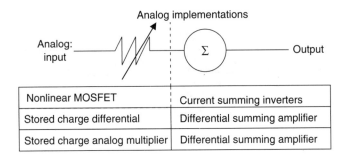

**FIGURE 7.22**
Analog implementations

In analog implementations, the synapse is implemented as a variable resistor and the neuron is implemented as a multi-input summing amplifier. The methods employed to implement these elements are almost as numerous as the number of designs. Figure 7.22 illustrates several representative ones.

The advantages of analog over digital implementations are: the analog circuitry is much smaller; the number of circuit elements required to implement an analog synapse or neuron is much lower and the connectivity between neurons requires a single path as opposed to multiple lines required in digital technology. Thus, a much greater packaging density may be achieved in analog implementations. Also, since the circuitry is simple, "on-chip training" is easier to achieve. Other advantages include direct sensor interface and higher throughput. On the negative side, analog implementations suffer from long-term memory instability, because of the charge leakage in synapse storage. Circuit fabrication and temperature sensitivity may also contribute to errors. Also, the input interface is inflexible since all inputs to a specific analog implementation must be conditioned and scaled to match the performance requirements of the design.

Hybrid implementations combine digital and analog circuitry in the same design. A set of digitally selected resistors (Fisher et al., 1991) or MOSFETs (Khachab and Ismail, 1991) are used to implement the synapse weighting, thus utilizing the digital advantage of reading and writing synaptic weights, combined with the greater density of analog designs.

Optical implementations are not as well developed as other approaches, yet they offer the greatest potential for future growth in ANN systems. Figure 7.23 illustrates an optical approach in which holograms are used to implement the synaptic function (9). In optical implementations connectivity is achieved via light. Since crossing light beams do not interfere, one of the greatest problems in implementing high density ANNs is eliminated. Optical ANNs have the potential of achieving the highest connectivity and density of all the approaches. Another advantage of optical ANNs is that the highest throughput is achieved. The greatest difficulty in practical implementation is interfacing to a variety of input types, and generation of holographic attenuators.

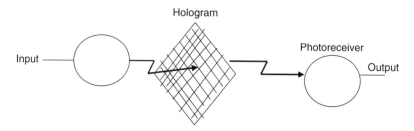

**FIGURE 7.23**
Optical implementations

Other implementations of ANN systems include simulations executed on SIMD and MIMD architectures (Iwata et al., 1989). While these approaches generally outperform simulations on conventional computer architectures, their throughput does not match that of hardware implementations.

### 7.3.1.7 A programmable analog neural network processor

This section is extracted from (Fisher et al., 1991) and provides a brief description of an analog neural network breadboard consisting of 256 neurons and 248 programmable synaptic weights. The breadboard has been used to demonstrate the application of neural network learning to the problem of real-time adaptive mirror control. The processor was designed to maximize its analog computational characteristics and flexibility. It can be used to study both feedforward and feedback networks or as a massively parallel analog control processor. It consists of an array of custom programmable resistor chips assembled at the board level.

The analog neural network processor shown in Figure 7.24 consists of 256 operational-amplifier-based neural circuits with a total of 2048 programmable resistors physically located on 16 boards (neuron boards). The resistors, which are programmed via a personal computer (PC), provide a means of training the analog neural network. The C programming language was used to write an extensive library of routines that can be used as part of specialized programs within an overall system shell. The input to the neuron boards can be from analog sources or from 12-bit digital sources converted to analog. The digital sources can be provided by the PC or from external sources. Thus it is possible to test an application using simulated data from the PC or from external sources. The outputs can be monitored with a set of 256 analog-to-digital converters.

The digitally programmable analog resistor chip was produced using a commercial analog gate array process. Each channel of the chip consists of a buffered source follower operational amplifier connected to five polysilicon resistors through bipolar analog switches. The values range from 8 to 128 k$\Omega$. Hundreds of these resistors to be connected in parallel without placing

# Current Directions

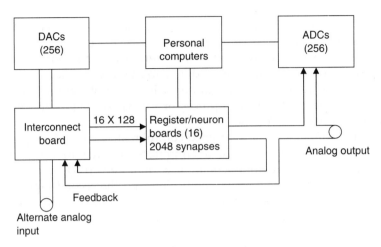

**FIGURE 7.24**
Analog network Processor (Fisher et al., 1991) (Reprinted from IEEE. With permission.)

too large a load on either the external summing amplifier or the input circuit. The data word is latched within the chip and hence no refresh is required.

The interconnection crossbar structure, the weights (or resistors), and the operational amplifiers are on a single board. The processor is on a separate board. This design approach allows the system to be expanded in a modular fashion; the complexity expands linearly with the number of required output neurons. The interconnect boards are problem specific, while the rest of the analog processor is general purpose.

The neuron board can be configured to have as few as seven inputs to each of the 16 neurons or as many as 128 inputs to a single neuron. One channel per resistor chip is used to program the feedback resistor. By using a programmable resistor from the package for the feedback resistance, the neuron gain becomes independent of the actual value of the internal package resistance. The gain is dependent only on the ratio of the resistances within the package. These can be controlled much closer than the package to package variation. The board may be setup as an integrator by installing a capacitor across the feedback resistor. The resistor (weights) on the board may be reprogrammed at a rate of one per microsecond.

The first application of the analog processor was the demonstration of neural network control of 21 actuators of a 69-actuator adaptive mirror. The purpose of such a system is to correct for distortions introduced when light forming the image passes through the atmosphere. Applications of this technology include the correction of solar and other bright celestial object images and the up-linking of laser beams through the atmosphere. The slope of the wavefront is measured at a large number of points and a best fit calculation is then done to determine the position of actuators which will then undo the distortions when the light making up the image reflects from the adaptive mirror surface. The system must operate with a closed-loop control bandwidth of at least 100 Hz to keep up with the atmospheric changes. This is equivalent to having a step response settling time of less than 10 msec.

The overall system includes a wavefront sensor which provides the input signals to the system through a set of digital to analog converters. The output of the neuron boards drives the integrators whose output is amplified to drive the adaptive mirror. A STAR array processor is used as a diagnostic to evaluate the performance of the system.

In order to implement the sparsened control matrix method, 16 programmable resistors and two operational amplifiers were required per actuator. Because the operational amplifiers are used in an inverting mode, the first takes inputs which have positive coefficients and the second inputs which have negative coefficients. In order to obtain a type one servo, the output of the neuron boards is fed to a set of integrators. A software controlled switch is used to open and close the control loop.

A laser and beam expander was used to direct a beam into the adaptive mirror. An image from the mirror reflects from the back side of a beam splitter onto the wavefront sensor. This beam is also combined with a reference beam to produce an interferogram image on a CCD array focal plane. The interferogram provides an independent, real-time diagnostic of the system performance.

There are two distinct stages of training. The first stage produces the best possible open-loop control coefficients, and the second modifies these control coefficients to obtain improved real-time closed-loop performance.

The best set of weights for a minimum time solution are the weights that are optimal for an open-loop correction or the correction that one gets if just one frame of data is used to compute the new actuator positions. This is the type of correction one would have if the network were simply a feedforward network. In order to obtain this set of weights an automated program was written that can supply a test voltage to each of the actuators at a time. During each actuator "punch" each of the resistors connected to the control of the actuator was turned on one at a time and the response was recorded. This set of measurements is a complete training set. While these data could be used with a traditional neural network training routine such as the LMS algorithm to obtain a control matrix, it is more straightforward to calculate the pseudo inverse of the data. Such an inversion is possible because the dataset is guaranteed to be complete and nondegenerate. Such is not the case for randomly sampled data typically used for training. The resulting control matrix did have contributions to non-nearest neighbor connections not implemented in the hardware. However, the largest contributions were at the nearest neighbors that had been prewired. Contributions to unwired elements of the control matrix were ignored. The resistor values were then scaled to be within the 5-bit range of the resistor chip. The gain-determining resistors were adjusted to maximize the closed-loop gain while avoiding oscillation on any of the 21 actuators. There was more than a factor of 3 margin on the range of gain from a critically damped signal with a damped oscillation with a decay time of less than 1 sec. Finally, the values of the resistors for the feedback connections were set to the value of the gain resistor divided by the number of the nearest neighbors. The values of these resistors are determined from the interconnectivity and cannot be larger without causing interactuator oscillations.

The first test of the flexibility of the system came when the optical alignment difficulties within the wavefront sensor limited the number of actuator measurements such that the control was valid only for the inner 21 actuators. Although the system had been wired for 69 actuators, only software modifications and changes in the control matrix values were required to reduce the analog control to the smaller number. This illustrates the advantage of the programmable resistor design.

#### 7.3.1.8 Applications

Neural networks have been used to match patterns, make generalizations, merge new situations into old experiences, mirror the structures in their environment, and find the best fit among many possibilities. In an application to grade textile yarns, a back propagation neural network (BNN) and Karhunen–Loeve (K–L) expansion method are used (Lien and Lee, 2004). The ensemble of artificial neural networks applicable in weather forecasting are contrasted in (Maqsood et al., 2004). Application of neural networks to human-like fault diagnosis are discussed in (Ariton and Palade, 2004).

### 7.3.2 DNA Computing

In 1980s scientists visualized producing a von Neumann type computer in carbon, rather than silicon (Conrad, 1985). That visualization has led more recently to DNA computing, one of the fastest growing fields in both Computer Science and Biology. DNA molecules encode genetic information for all living things. In DNA computers, the problem is encoded in the molecules present in a suspended solution of DNA, and certain combinations of DNA molecules are interpreted as particular solutions to the problem. Thus, the information specifying a computational problem too complex for even a supercomputer can be encoded in DNA and using various molecular-biological tools (enzymes) the information is processed.

The key advantage of DNAs is that it is massively parallel in nature. More than 10 trillion DNA molecules can fit into an area no larger than 1 cubic centimeter . With this small amount of DNA, a computer would be able to hold 10 TB of data, and perform 10 trillion calculations at a time. The large supply of DNA makes it a cheap resource and they are many times smaller and light weight than silicon structures used in today's computers. Unlike the toxic materials used to make silicon processors, DNA biochips can be made cleanly. DNA computers consume low power and can be used for solving complex problems.

However, DNA computing is very slow since each primitive operation takes hours to run in a test tube of DNA molecules. All the algorithms proposed so far use slow molecular-biological operations. DNA molecules fracture and gradually turn into water with time (Hydrolysis) and it is hard to control the operations performed by the DNA computer. As such, they are very unreliable. While the DNA computers can perform massively parallel operations, transmitting information from one molecule to another is a problem. Thus, as of now DNA computing is not fully practical.

Just like a string of binary data is encoded with ones and zeros, a strand of DNA is encoded with four *bases*, represented by the letters A, T, C, and G. The bases (also known as *nucleotides*) are spaced every 0.35 nm along the DNA molecule, giving DNA a data density of nearly 18 Mbits inch$^{-1}$. Another important property of DNA is its double-stranded nature. The base A binds with T and the base C binds with G to form base pairs. For example if the sequence S is: ACAATG then its complement S' would be TGTTAC. This complementary nature of DNA is exploited and is used for Error correction. Sometimes, DNA enzymes make mistakes by cutting DNA strands where they should not, or inserting a T instead of G. DNA can also be damaged by thermal energy and UV energy from the sun. If the error occurs in one of the strands of double-stranded DNA, repair enzymes can restore the proper DNA sequence by using the complement strand as a reference.

Leonard Adleman, often called the inventor of DNA computers introduced the idea of DNA to solve various complex problems in 1994. He solved the directed Hamiltonian Path Problem (an NP complete problem) popularly known as Travelling Salesman Problem. Given a directed graph, where each node represents the city, the goal of it is to find the shortest route between two cities going through each city only once.

Adleman chose to find the shortest route between seven cities. He took the DNA strands to represent the seven cities. Each sequence of A, T, G, and C represented the cities and possible flight path. He then mixed the molecules in the test tube, with some DNA strands sticking together. A chain of these strands represented a possible answer. When all the possible combinations of DNA strands were formed that represent all the possible routes between cities, he used polymerase chain reaction technique to produce many copies of the specific sequence of DNA He used a method known as "Gel Electrophoresis" to find the final solution. The basic principle behind Gel Electrophoresis is to force DNA through a gel matrix by using an electric field. DNA is a negatively charged molecule under most conditions, so if placed in an electric field it will be attracted to the positive potential. However, since the charge density of DNA is constant, long pieces of DNA move as fast as short pieces when suspended in a fluid. Thus by using a gel, the longer pieces are forced to slow down at different rates depending on their length. Thus the one that represents the correct solution, connecting all the cities was isolated. This experiment marked the beginning of DNA computing. Computer Scientists since then have been trying to find ways of using the power of DNA strands to be useful for DNA computing.

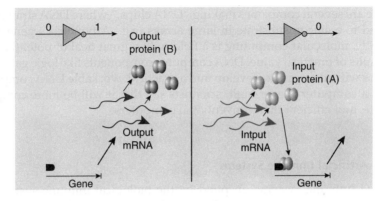

**FIGURE 7.25**
Inverse logic (http://www.princeton.edu/~rweiss/bio-programming/weiss-natural-computing-2003.pdf)

Weiss and his colleagues from Princeton University have been able to analyze a new genetic signal processing circuit that can be configured to detect various chemical concentration ranges of ligand molecules. These molecules freely diffuse from the environment into the cell. Figure 7.25 shows the inverse logic. A digital inverter consists of a gene encoding the instructions for protein B and containing a region (P) to which protein A binds. When A is absent (left) — a situation representing the input bit 0 — the gene is active, and B is formed — corresponding to an output bit 1. When A is produced (right) — making the input bit 1 — it binds to P and blocks the action of the gene — preventing B from being formed and making the output bit 0. For further details refer to Weiss et al. (2003).

Researchers at the Weizmann Institute of Science have devised a computer that can perform 330 trillion operations per second, more than 100,000 times the speed of the fastest PC. It runs on DNA and is used to identify changes in the body's balance of molecules indicating the presence of certain cancers, to diagnose the type of cancer, and to react by producing a drug molecule to fight the cancer cells. It consists of chains consisting of three main segments. The first segment is used to sense levels of substances which are produced by cancerous cells. It functions like a computer running through a simple algorithm. One algorithm is intended to diagnose prostate cancer. It infers that if levels of two messenger RNA molecules are lower than usual, and levels of two others are elevated, there must be prostate cancer cells in the vicinity.

If the analytical/computational segment decides that cancer is present, it tells the second segment to release the third segment, which is an anti-cancer drug — in this case, consisting of so-called anti-sense DNA. Thus in one series of test-tube experiments, the computer was programmed to identify RNA molecules that indicate the presence of prostate cancer and following a correct diagnosis, to release the short DNA strands designed to kill cancer cells. Similarly, one form of lung cancer was also identified.

There are several companies making "DNA chips," where DNA strands are attached to a silicon substrate in large arrays (e.g., Affymetrix's genechip). Currently, molecular computing is a field with a great deal of potential, but few results of practical value. DNA computer components like logic gates and biochips will take years to develop into a practical, workable DNA computer. If such a computer is ever built, scientists say that it will be more compact, accurate, and efficient than conventional computers.

### 7.3.3 Artificial Immune Systems

Artificial Immune Systems, inspired by the biological immune system are now an area of active research to build systems that can solve various computational problems (Dasgupta, 1999). The human immune system exhibits many useful computational aspects like recognition, memory, and learning. The main aim of the human immune system is to defend the body from foreign invaders like viruses and bacteria and categorize them as "self" and "nonself."

A huge number of immune cells circulate throughout the human body. One of the main immune cells participating in immune response is the *lymphocyte*. There are two main types of lymphocytes that make up the immune system: T-cells and B-cells. These two types play different roles in immune response. The lymphoid system consists of primary organs (bone marrow, thymus gland) and secondary organs (at or near possible portals of pathogens like adenoids, spleen, tonsils etc.). Different populations of lymphocytes circulate at primary and secondary lymphoid organs such that appropriate populations of B- and T-cells are recruited to different locations in the body.

Various features of the immune system can be utilized in forming artificial immune systems that make it possible to process information. The very first feature is the capability to recognize and classify the self and nonself cells and generate the selective response.

Whenever the lymphocytes are activated, a few of each kind become special memory cells which are content-addressable. The inherent longevity of immune memory cells is dynamic and requires continued stimulation by antigens. This memory then allows remembering the structure of the specific antigen so that any subsequent attack from the same type of antigen leads to rapid immune response. The immune system generates diverse set of lymphocytes receptors and ensures that at least some of them bind to the given antigen.

The other computational features of the immune system are learning and feature extraction. They learn the structure of the antigen by experience and change the concentration in the lymphocyte accordingly. Antigen presenting cells (APC) capture and process antigens for presentation to T-lymphocytes. Each APC serves as a filter that destroys the molecular noise and a lens that focuses the attention of the lymphocyte-receptors. There is no central organ that controls the functions of the immune system. The immune cells are "self regulatory." They are distributed and circulated to all parts of the body through the blood, lymph, lymphoid organs, and tissue spaces. This helps in "distributed detection." Since lymphocytes can bind to various structurally related antigens, the cross reaction in immune system is stochastic in nature. Whenever the matching to the antigen exceeds the threshold limit, immune system responds and binds to the antigen. Finally, the process called clonal expansion generates a large population of antibody-producing cells that are specific to the antigen. It results in destroying or neutralizing the antigen.

There are various differences between neural networks and artificial immune system (AIS). The neural network uses the neurons which are approximately $10^{10}$ in number whereas immune cells (lymphocytes) are used in AIS which are of the order of $10^{12}$. Neurons have fixed position while lymphocytes are distributed throughout the body. Neurons are considered to be homogeneous while immune cells are heterogeneous agents. The neural network needs to distinguish between the stored patterns and conscious imagination whereas AIS needs to distinguish between the "self" and the "nonself" cells. As is already mentioned, immune systems are self-regulatory and there is no central organ regulating them. On the other hand, brain controls the functioning of the nervous system. Communication between the neurons are through linkage hardware like axons, dendrites as electrical signals while the communication in immune cells takes place through transient cell–cell contacts and via secretion of soluble molecules.

Some of the similarities between the two systems are: both the systems have memory and learn with experience. Synaptic strength patterns in neurons constitute memory which is auto-associative and nonhereditary. On activation, lymphocytes become special memory which are content addressable but not hereditary. Both the systems are robust. Neural network is flexible and damage-tolerant while immune system is scalable and self-tolerant. Both neurons and immune cells are fired after a threshold value.

There are various models constructed based on the immune system principles. One of the most common is the negative selection algorithm (Forrest et al., 1994), based on the fact that the T-cells have receptors on their surface that can detect antigens. These help in the discrimination against self and nonself in immune system. During the generation of T-cells, receptors are made by a pseudo-random genetic rearrangement process. Then they undergo a censoring process called negative selection in the thymus where T-cells that react against self-proteins are destroyed, so only those that do not bind to self-proteins are allowed to leave the thymus. These matured T-cells then circulate throughout the body to perform immunological functions to protect against foreign antigens.

The negative selection is simple and works on the same principle. It first defines a self-set S which may be a data file or a program that needs to be protected. It then generates the set of detectors R which are pattern recognizers and do not match any string in S. It uses partial matching rule instead of perfect matching and match if and only if they are identical in at least r contiguous positions, where r is a suitably chosen parameter. It monitors matching of detectors R against S. If any detector ever matches, then a change is known to have occurred, because the detectors are designed to not match any of the original strings in S.

The detectors R are generated by an iterative process. First the strings are randomly generated and placed in set P. The affinity of all strings of the self-set S are then generated. If the affinity of all strings in P with at least one string of S is greater than or equal to a given affinity threshold, then the string in P recognizes the self-string and has to be eliminated (negatively selected) otherwise string in P belongs to nonself set and is included in set R. Thus detectors are generated and those that detect self are eliminated, so that the remaining detectors can detect nonself. As the size of the self grows, the generation of the detectors becomes increasingly difficult.

The negative selection algorithm has been used in various applications. One of the applications is Virus Detection (Forrest et al., 1994) where data and program files were protected against different viruses. A number of experiments were first conducted in DOS environment with different viruses. The negative selection algorithm was then used to detect the variations of these viruses and the results showed that this method could detect the modification that occurred in data files due to virus infection. There are various advantages of using this method. It provides a high system wide reliability at low cost. It is probabilistic and tunable and learns from its previous experience. It can detect the new viruses which have not been identified previously. But the major problem of this method is that it can only be used to protect static data files of software. Another application of negative selection algorithm is in the UNIX process monitoring (Forrest et al., 1996) to detect the computer system for the harmful intrusion.

Artificial immune systems are currently an active area of research interest and no practical computer system based on these principles has been built.

## 7.4 Optical Computing

Optical computing has been an active area of research over the last few years, as an alternative to electronic digital computing to design high-speed computer systems. Data communication through optical fibers has already been proven to be more versatile and cost-effective than that through the electronic media. Optical interconnections and optical integrated circuits provide a way out of these limitations to computational speed and complexity inherent in conventional electronics. During the last decade, there has been continuous emphasis on the following aspects of optical computing:

- Optical tunnel devices are under continuous development varying from small caliber endoscopes to character recognition systems with multiple type capability.
- Optical processor development for asynchronous transfer mode.

- Development architectures for optical neural networks.
- Development of high accuracy analog optical processors, capable of processing large amounts of data in parallel.

Scientists at Bell Labs (Dodabalapur, 1998) have used polymer FETs and LEDs, made from various organic materials, one of which allows electrons to flow, another which acts as highway for holes (the absence of electrons); light is produced when electrons and holes meet and have come up with a 300-micron-wide pixels. The pixels have the brightness of about 2300 candela $m^{-2}$.

Recall that the speed characteristics of electronic computational structures are functions of device speeds and the architecture. There is a large disparity between the speed of the fastest electronic switching component and the speed of the fastest digital electronic computers. The switching speeds of transistors are as high as 5 psec, while the fastest computers operate at clock periods of the order of a few nanoseconds. The limitations of electronic technology that cause this speed disparity are (Jordan, 1988): electromagnetic interference at high speeds, distorted edge transitions, complexity of metal connections, drive requirements for pins, large peak power levels, and impedance matching effects.

Electromagnetic interference is the result of coupling of the inductances of two current carrying wires. Sharp edge transitions are a requirement for proper switching. But higher frequencies attenuate greater than lower frequencies, resulting in edge distortions at high speeds. The complexity of metal connections on chips, circuit boards, and between system components introduces complex fields and unequal path delays. The signal skews introduced by unequal path delays are overcome by slowing the system clock. Large peak power levels are needed to overcome residual capacitances. Impedance matching effects at connections require high currents and in turn cause lower system speeds. There are several advantages to using free-space optics for interconnections (Jordan, 1988). By imaging a large array of light beams onto an array of optical logic devices, it is possible to achieve high connectivity. Since physical interconnects are not needed (unless fibers or waveguides are used) connection complexity and drive requirements are reduced. Optical signals do not interact in free space (i.e., beams can pass through each other without any interference) and hence a high bandwidth can be achieved. Moreover, it is free from electrical short circuits. There is no feedback to the power source as in electronic circuits and hence there are no data dependent loads. The inherently low signal dispersion of optical signals implies that the shape of a pulse as it leaves its source remains virtually unchanged until its destination. Another advantage of optics over electronics is communication. Electronic communication along wires requires charging of a capacitor that depends on length. In contrast, optical signals in optical fibers, optical integrated circuits, and free space do not have to charge a capacitor and are therefore faster. Optical devices can be oriented normal to the surface of an optical chip such that light beams travel in parallel between arrays of optical logic devices rather than through pins at the edges of chips as in electronic integrated circuits. Lenses, prisms, and mirrors can convey an image with millions of resolvable points in parallel. Thus, optical data processing can be done much easier and less expensive in parallel than can be done in electronics.

It is important to note the differences in basic characteristics of electrons and photons. Electrons easily affect each other even at a distance, thus making it easy to perform switching. But, this ease of interaction complicates the task of communication, since the signals must be preserved. Since it is very difficult to get two photons to interact, it is very difficult to get two optical signals to interact. Thus, optics is bad for switching but good for communications. A solution may be to stay with hybrid technology where electronics performs all the computations and the optics performs all the communication. Moreover, since photons are uncharged and do not interact with one another as readily as electrons. Light beams may pass through one another in full duplex operation, for example, without distorting the information carried. In the case of electronics, loops usually generate noise voltage spikes whenever the electromagnetic fields through the loop changes. Further, high frequency or fast switching pulses will cause interference in neighboring wires. Signals in adjacent fibers or in optical integrated channels do not affect one another nor do they pick up noise due to loops. Finally, optical materials possess superior storage density and accessibility over magnetic materials.

There are several problems associated with the optical technology. Most of these problems stem from an inability of optical signals to interact and thus perform switching. Electronics technology is mature, cost-effective, and allows the fabrication of high-density switching components. Photonics on the other hand, is less mature and requires tight imaging tolerances and constant power consumption for modulator-based optical devices. Optical devices can be spaced a few micron apart on optical chips but require several centimeters of interaction distance for lenses, gratings, and other imaging components. Micro-optic techniques are being investigated as solutions to this problem.

A number of optical devices (logic gates, optical switches, optical interconnections, and optical memory) have been manufactured. Also, optical CD-ROM discs are now very common in home and office computers. Scalable crossbar systems now use free-space optical techniques for the arbitrary interconnections between a set of inputs and a set of outputs. Optical sorting and optical crossbar interconnects are used in asynchronous transfer modes or packet routing and in shared-memory multiprocessor systems. All these devices and technologies ultimately lead to real optical computers.

Many approaches have been proposed for forming general-purpose optical computers. A group of researchers from the University of Southern California, jointly with a team from the University of California, Los Anglos, have been able to develop an organic polymer with a switching frequency of 60 GHz. This is three times faster than the current industry standard, lithium niobate crystal-based devices. Another group at Brown University and the IBM Almaden Research Center (San Jose, CA) has achieved ultra fast switching down to 100 psec. They have used ultra fast laser pulses to build ultra fast data storage devices. Their results are almost ten times faster than currently available "speed limits."

To develop a method for interconnecting circuit boards optically NEC (Tokyo, Japan) has used Vertical Cavity Surface Emitting Laser arrays (VCSEL). Researchers at Osaka City University (Osaka, Japan) reported using a set of optical fibers to automatically align a set of optical beams in space. Researchers at NTT (Tokyo, Japan) have achieved 1000 interconnections per printed-circuit board, with throughput ranging from 1 to 10 TB $sec^{-1}$. They designed an optical back plane with free-space optical interconnects using tunable beam deflectors and a mirror.

The most recent research is at the NASA/MSFC laboratories. Abdeldayem et al. (1999) have designed an optical logic gate. It is a switch that controls one light beam by another; it is "ON" when the device transmits light and is "OFF" when it blocks the light. Two fast all-optical switches using Phthalocyanine thin films and Polydiacetylene fiber have been demonstrated in laboratory at NASA/Marshall Space Flight Center. The phthalocyanine switch is in the nanosecond regime and functions as an all-optical AND logic gate, while the polydiacetylene one is in the picosecond regime and exhibits a partial all-optical NAND logic gate. In the Nanosecond All-Optical AND-Logic gate, two focused collinear beams were waveguided through a thin film (~1 mm thickness and few millimeters in length) of metal-free phthalocyanine. To demonstrate the AND gate in the phthalocyanine film, two focused collinear beams through a thin film of metal-free phthalocyanine film were waveguided. The film thickness was and a few millimeters in length. The two beams, one from the pulsed Nd:YAG laser with pulse duration of 8 nsec at 532 nm and the other, CW He–Ne laser at 633 nm were used. They were then focused by a microscopic objective and sent through the phthalocyanine film. A narrow band filter then blocks the 532 nm beam and allows the He–Ne beam to pass through at the output. This transmitted beam was then focused on a fast photo-detector and to a 500 MHz oscilloscope. It was found that the transmitted He–Ne CW beam was pulsating with a nanosecond duration and in synchronous with the input Nd:YAG nanosecond pulse. The setup described above demonstrated the characteristic table of an AND logic gate. A schematic of the setup is shown in Figure 7.26.

All-Optical NAND logic gate setup is similar to AND logic gate setup except a hollow fiber filled with a polydiacetylene is used instead of phthalocyanine film. Diacetylene monomer was injected into the hollow fiber and polymerized by Ultra-Violet (UV) lamps to prepare polydiacetylene fiber. The two beams from the Nd:YAG Pulsed Laser of 532 nm and CW He–Ne laser of 633 nm were focused onto one end of the fiber. At the other end of the fiber a lens was focusing the output onto the narrow slit of a monochrometer with its grating set at 632.8 nm. A fast detector was attached to the monochrometer and sending the signal to a 20 GHz digital oscilloscope. It was found

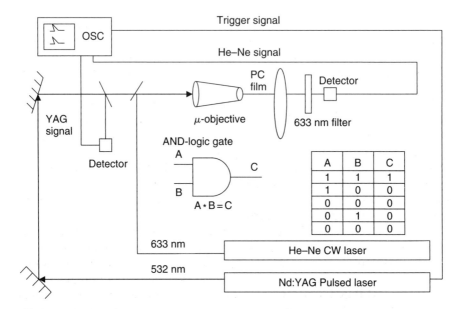

**FIGURE 7.26**
A schematic of the nanosecond all-optical AND logic gate setup (Abdeldayem et al., 1999)

that Nd:YAG pulse induced a weak fluorescent picosecond signal (40 psec) at 632.8 nm with the He–Ne beam OFF. When the He–Ne beam was turned on, this signal disappeared each time. These results exhibit a picosecond respond in the system and demonstrated three of the four characteristics of a NAND logic gate as shown in Figure 7.27.

Optical computers use photons traveling on optical fibers or thin films instead of electrons to perform the appropriate functions. Lenslet, an Israeli company has developed an optical computer "EnLight" that can perform 8000 billion arithmetic operations per second. EnLight is a hybrid device that combines both electronic and optical circuits. EnLight is being introduced to the digital signal processor (DSP) market. Smart Antennas that can sense the environment they are operating in are being produced using EnLight. These antennas optimize the way they transmit and receive information.

One of the available vectors DSP with an embedded optical core is Lenslet's EnLight256 DSP. It provides 125 Mega vector–matrix multiplications and 500 Mega vector–vector operations per second on matrices of size 256 × 256 and vectors of size 256 elements. The EnLight256 architecture consists of several processing layers. The optical core performs a 256 8-bit element vector by a 256 × 256 8-bit element matrix multiplication in a single clock cycle. Vector–vector and scalar processing are also required in order to implement a complete algorithm. Three main tools are used to develop the software on EnLight256. They are: MATLAB APL bit exact stimulator, APL Studio bit exact simulator, and the APL Studio Emulator.

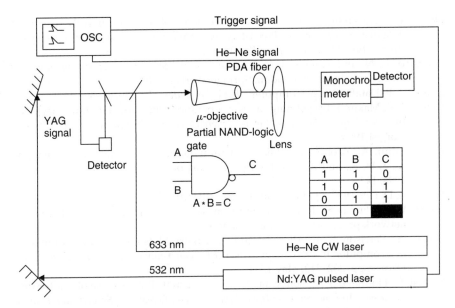

**FIGURE 7.27**
A schematic of the all-optical NAND logic gate setup (Abdeldayem et al., 1999.)

For further details on optical computing refer to the book by Murdocca (1990), and excellent surveys on various aspects of optical computing in (Goswami, 2003; Pramono and Endarko, 2001; Yu and Jutamulia, 2000).

## References

Abdeldayem, H., Frazier, D., Paley, M., and Witherow, W. K. Recent Advances in Photonic Devices for Optical Computing, NASA Marshall Space Flight Center, Space Sciences Laboratory Report, 1999.
Ackerman, W. B. Dataflow languages. *IEEE Computer*, 15(2), 1982, 15–25.
Affymetrix Genechip, http://affymetrix.com/
Agranat, A. J., Neugebaur, C. F., Nelson, R. D., and Yariv, A. The CCD neural processor: A neural network integrated circuit with 65536 programmable analog synapses. *IEEE Transactions on Circuits and Systems*, 37, 1990, 246–253.
Ariton, V. and Palade, V. Human-like fault diagnosis using a neural network implementation of plausibility and relevance. *Neural Computing and Application*. London: Springer-Verlag, 2004.
Arvind and Nikhil, R. S. Can dataflow subsume von Neumann computing? *Proceedings of the ACM*, 1989, 262–272.
Conrad, M. On design principles for a molecular computer. *Communications of the ACM*, 28, 1985, 464–480.
Dasgupta, D. *Artificial Immune Systems and Their Applications*. Berlin: Springer-Verlag, 1999.

Davis, A. L. The architecture and system method of DDM1: a recursively structured data driven machine. In *Proceedings of the 5th Annual Symposium on Computer Architecture*, New York, pp. 210–215, 1978.

deCastro, L. N. and Timmis, J. I. *Artificial Immune Systems: A New Computational Intelligence Approach*. London: Springer-Verlag, 2002.

Dennis, J. B. Dataflow supercomputers. *IEEE Computer*, 13, 1980, 48–56.

Dennis J. B. Maximum pipelining of array operations on static data flow machines. In *Proceedings of the International Conference on Parallel Processing*, pp. 176–184, August 1983.

Dodabalapur, A., Bao, Z., Makhija, A., Laquindanum, J. G., Raju, V. R., Feng, Y., Katz, H. E., and Rogers, J. Organic smart pixels. *Applied Physics Letters*, 73, 1998, 142.

Feitelson, D. *Optical Computing*. Cambridge, MA: MIT Press, 1988.

Fisher, W. A., Fujimoto, R. J., and Smithson, R. C. A programmable analog neural network processor. *IEEE Transactions on Neural Networks*, 2, 1991.

Forrest, S., Perelson, S. A., Allen, L., and Cherukuri, M. Self–nonself discrimination in a computer. In *Proceedings of the IEEE Symposium on Research in Security and Privacy*, pp. 202–212, May 1994.

Forrest, S., Hofmeyr, S. A., Somayaji, A., and Longstaff, T. A. A sense of self for unix processes. In *Proceedings of IEEE Symposium on Research in Security and Privacy*, pp. 76–80, 1996.

Foster, I. and Kesselman, C. *The GRID2: Blueprint for a New Computing Infrastructure*. San Francisco, CA: Morgan Kaufmann, 2004.

Foster, I., Kesselman, C., and Tuecke, S. The anatomy of the grid: enabling scalable virtual organizations. *International Journal of Supercomputer Applications*, 15, 2001, 200–222.

Gaudiot, J. and Bic, L. *Advanced Topics in Data-Flow Computing*. Englewood Cliffs, NJ: Prentice Hall, 1989.

Glauert, J. A single assignment language for dataflow computing, Master's Thesis, University of Manchester, Manchester, UK, 1978.

Goswami, D., Optical computing, optical components and storage systems. *Resonance*, pp. 56–71, June, 2003.

Grid Computing, *The DoD SoftwareTech News*, April 2004.

Hecht-Nielson, R. Performance limits of optical, electro-optical, and electronic neurocomputers. *Optical and Hybrid Computing*, 634, 1986, 277–306.

Hicks, J., Chiou, D., Ang, B., and Arvind. Performance studies of the monsoon dataflow processor. *CSG Memo 345-2*, MIT, October 1992.

Hollis, P. W. and Paulos, J. J. Artificial neural networks Using MOS analog multipliers. *IEEE Journal of Solid State Circuits*, 25, 256–272, 1990.

Hopfield, J. J. and Tank, D. W. Computing with neural circuits: a model. *Science*, 233, 1986, 625–633.

Iwata, A., Sato, Y., Suzumura, N., Matsuda, S., and Yoshida, Y. An Electronic Neurocomputer Using General Purpose Floating-Point Digital Signal Processors, Department of Electrical and Computer Engineering, Nagoya Institute of Technology, Showa-ku, Nagoya, Japan, 1989.

Jewell, J. L., Murdocca, M. J., McCall, S. L., Lee, Y. H., and Scherer, A. Digital optical computing: devices, systems, architectures. In *Proceedings of the Seventh International Conference on Integrated Optics and Optical Fiber Communication*, Kobe, Japan, pp. 147–156, July 18, 1989.

Jordan, H. F. Report of the Workshop on All-Optical, Stored Program, Digital Computers, Technical report, Department of Electrical and Computer Engineering, University of Colorado at Boulder, 1988.

Khachab, N. L. and Ismail, M. A New Continuous-Time MOS Implementation of Feedback Neural Networks, Solid-State Laboratory, Ohio State University, 1991.

Kirkham, C. C., Watson, I., and Gurd, J. R. The Manchester prototype dataflow computer. *Communications of the ACM*, 28, 1985, 34–52.

Lenslet, EnLight256, 8000 Giga MAC/sec fixed point DSP, http://www.lenslet.com/docs/EnLight256_White_Paper.pdf

Lien, H. C. and Lee, S. Applications of Neural Networks for Garding Textile Yarns. *Neural Computing and Application*. London: Springer-Verlag, 2004.

Maqsood, I., Md. Khan, R., and Abraham, A. *An Ensemble of Neural Networks for Weather Forecasting*. *Neural Computing and Application*. London: Springer-Verlag, 2004.

Murdocca, M. *A Digital Design Methodology for Optical Computing*. Boston, MA: MIT Press, 1990.

Pinfold, W. Meiko scientific's computing surface: a highly scalable parallel processor for $C^3I$. *Military and Aerospace Electronics*, 2, 1991, 37–38.

Pramono, Y. H. and Endarko, M. Nonlinear waveguides for optical logic and computation. *Journal of Nonlinear Optical, Physics and Materials*, 10, 2001, 209–222.

Robinson, T. Mike Hochberg, and Steve Renals. The Use of Recurrent Neural Networks in Continuous Speech Recognition, Cambridge University Engineering Department, UK, 1995, http://mi.eng.cam.ac.uk/~ajr/rnn4csr94/rnn4csr94.html

Rumbaugh, J. A dataflow multiprocessor. *IEEE Transactions on Computers*, C-26, 1977, 1087–1095.

Ryu, W. DNA Computing: A Primer, http://arstechnica.com/reviews/2q00/dna/dna-1.html

Sharp, J. A. *Dataflow Computing*. New York: John Wiley & Sons, 1985.

Shiva, S. G. *Computer Design and Architecture*. New York: Marcel-Dekker, 2000.

Stefan Lovgren. Computer made from DNA and enzymes. *National Geographic News*, February 24, 2003, http://news.nationalgeographic.com/news/2003/02/0224-030224-DNAcomputer.html.

Tanenbaum, A. S. *Computer Networks*, New York: Prentice-Hall, 1991.

Tesler, L. G. and Enea, H. J. A language design for concurrent processes. In *Proceedings of AFIPS Spring Joint Computer Conference*, Vol. 32, pp. 403–408, 1968.

The Weizmann Institute Genome Committee, Biological Computer Diagnoses Cancer and Produces Drug—in a Test Tube, 2004, http://www.bridgesforpeace.com/modules.php?name=News&file=article&sid=1666.

Treleaven, P. C. and Lima, I. G. Japan's fifth generation computer systems. *IEEE Computer*, 15, 1982, 79–88.

Yagi, T., Funahashi, Y., and Ariki, F. Dynamic model of dual layer neural network for vertebrate retina. In *Proceedings of the IEEE International Joint Conference on Neural Networks*, pp. 187–194, 1989.

Weiss, R., Basu, S., Kalmbach A., Karig, D., Mehreja, R., and Netravali, I. Genetic circuit building blocks for cellular computation. *Communications and Signal Processing*, 2003. Available at http://www.princeton.edu/!rweiss/bio-programming/weiss-natural-computing-2003.pdf.

Will Ryu. DNA Computing: A Primer, Arstechnica, 2004. Available at http://arstechnica.com/reviews/2q00/dna/dna-1.html

Yu, T. J. and Jutamulia, S. *Optical Signal Processing, Computing and Neural Networks*. Melbourne, FL: Krieger, 2000.

Zarri, G. A fifth generation approach to fifth generation intelligent information retrieval. In *Proceedings of the ACM Annual Conference*, p. 30, 1984.

# Index

Access time, 20
Advanced Load Address Table(ALAT), 32
Amdhal's law, 145
Application characteristics, 50
Architecture
    control-driven (control-flow), 55, 285
    data parallel, 60, 167
    data-driven (data-flow), 57, 285
    demand-driven (reduction), 57
    loosely coupled, 63
    load/store, 44, 100
    memory/memory, 23
    message passing, 226
    micro-architecture, 7
    MIMD, 60
    MISD, 58
    Non Uniform Memory (NUMA), 226
    parallel processing, 2, 27
    shared memory, 224
    SISD, 58
    SIMD, 59
    System, 10
    taxonomy, 55
    tightly-coupled, 63, 225
    uniform memory (UMA), 225
    Uniprocessor, 15
    VLIW, 119
Arithmetic/Logical Processors, 170
Arithmetic Logic Unit (ALU), 15
Arithmetic pipeline, 81
Array processor, 167
Artificial Immune System, 320
Asynchronous pipeline, 75
Average or sustained rate, 65
Axon, 306

Bandwidth, 5, 20, 183
Banking, 22
Barrier synchronization, 250
Back propogation, 307
Benchmarks, 65
    dhrystone, 67
    kernel, 66
    Lawrence Livermore loops, 66
    local, 66
    parallel, 68
    partial, 66

PERFECT, 68
Real world/Application, 66
    single processor, 66
SPECmarks, 67
SLALOM, 68
Stanford small programs, 68
synthetic, 67
Unix utility and application, 66
whetstones, 66
Benes Network, 240
Binary tree network, 189
Biology inspired computing, 305
Bitonic, 199
Block Transfer Engine, 273
Branch history, 105
Branch-prediction, 102
Burroughs Scientific Processor (BSP), 170
Bus
    structure, 16
    networks, 193, 235
    snooping, 231
    watching, 231
    window, 235
    width, 27

C*, 216
Cache, 11
    coherence, 229
    flushing, 231
    incoherency, 229
Cells, 287
Chaining, 139
Circuit switching, 186
CISC, 56
Cluster Grid (Departmental Computing), 301
CM-2, 207
CM-5, 268
COMPAS, 155
Completion rate, 126
Conditional branches, 102
Control Data Corporation (CDC)
    6600, 108
    STAR-100, 114
Control hazards, 80
Coarse-grained, 51
Collision vectors, 88

*331*

Communication bound, 54
Compilers, 9
Completely serial, 53
Computer network, 63
Computing paradigms, 2
Connectivity, 275
Control, 137
   processor, 169
   unit, 15
   driven (control flow) architectures, 55, 285
Cooperative autonomy, 299
Cost, 184
   factor, 68
   optimal, 202
Cray Research Corporation
   X1, 160
   T3D, 270
   X-MP CPU, 133

Data
   alignment network, 169
   driven architecture (data flow), 57
   dependencies, 52
   forwarding, 98
   hazards, 97
   interlocks, 96
   parallel architectures, 60, 167
   parallelism, 123
   speculation, 30
Dataflow
   architecture, 284
   Graphs (DFG), 287
   languages, 292
   models, 285
Deadlock, 251
Decentralized transfer, 183
Decode history table, 105
Dedicated paths, 183
Degree of parallelism, 49
Delayed branching, 103
Demand-driven (reduction) architectures, 55, 285
Dendrite, 305
Dhrystone, 67
Direct Memory Access (DMA), 26
Digital Signal Processor (DSP), 326
Distributed Management Task Force (DMTF), 302
Distributed processing systems, 299
DNA computing, 317
DOALL, 259
Dynamic pipeline, 107
Dynamic scheduling, 254
Dynamic topologies, 185

Easy (or trivial) parallelism, 54
Efficiency, 201
Enterprise Grid (enterprise computing), 301
Evolutionary approach, 283

Fat tree, 236
Feed forward network, 307
Feng's Taxonomy, 185
FETCH_AND_ADD, 251
Fine-grained, 51
Fork, 257
Flow control digits or FLITS, 186
Flynn's architecture classification, 58
Flynn's taxonomy, 55
Forbidden set, 90
Front-end processors, 27

Generality, 65
Granularity, 51
Grid computing, 98, 299
Global Grid (Internet computing), 302
Global Grid Forum (GGF), 302

Harvard architecture, 18
Hardware complexity, 184
Heavy-weight processes, 252
High level language (HLL) architectures, 57
High order interleaving (banking), 22
Highly scalable, 263
Hitachi Super Technical servers, 152
Hypercube, 211

Immune System, 320
Internal Forwarding, 99
International Business Machines (IBM)
   X, 274
   360/91, 110
Interrupts, 107
ILLIAC-IV, 205
Indirect transfer, 183
Instruction deferral, 108
Instruction pipeline, 119
Intel Corporation Itanium, 29
Interconnection networks, 353
   Benes, 240
   binary tree, 189
   bus, 236
   complete interconnection, 190
   connectivity, 183
   cost, 184
   crossbar FX/8, 193
   dynamic, 185
   fat tree, 236
   hypercube, 238
   mesh, 237

# Index

333

hardware complexity, 184
latency, 5, 114, 183
multistage, 196
perfect shuffle, 194
regularity, 184
ring(loop), 187
star, 188
static, 185
structures, 27
topologies
I-Structures, 297
Interleaving
   high-order, 22
   low-order, 21
Inverse logic, 319
Input/Output (I/O)
   channels, 27
   direct memory access (DMA) mode, 27
   front-end processors, 26
   processors, 26
   interrupt mode, 27
   Nodes(ION), 230
   programmed, 26
   subsystem, 26
Irregular topologies, 185

Join, 257

Latency, 5, 135, 183
Latency sequence, 135
Lawrence Livermore loops, 66
Level of parallelism, 51
Light-weight processes, 253
Linear pipelines, 82
Linpack, 66
*Lisp, 215
Load/store architecture, 44, 100
Local benchmarks, 66
Lock/Unlock, 248
Loop
   distribution, 151
   fusion, 151
   jamming, 151
   unrolling, 150
Loosely coupled, 63, 226
Low order interleaving, 21

Manchester data flow machine (MDM), 296
Memory, 34
   architecture, 23
   design considerations, 128
   constrained scaling, 262
   oriented vector processor architecture, 127
   technologies, 7

Message passing architecture, 226
Micro-architecture, 7
Micro-programmed control units, 24
Microtasking, 253
MIMD, 224
MIMD/SIMD architectures, 60
Minimum Average Latency (MAL), 135
MIPS Computer Systems R10000, 38
MISD, 58
MIT static architecture, 295
MIT/Motorola Monsoon system, 298
Multifunction pipeline, 84
Multilayered perceptron networks, 307
Multiple instruction buffers, 106
Multiport, 23
Multistage networks, 238

Neural networks, 305
Neuron, 305
NEWS grid, 212
NEC SX-6, 158
Nexus, 211
Node firing, 289
Nucleotides, 318
NUMA (non uniform memory access), 226

OASIS, 302
Open Grid Service Architecture (OGSA), 302
Omega Network, 240
One dimensional Networks, 187
Openness, 65
Operating system considerations, 242
Optical computing, 322
Optimal schedule, 254
Ownership protocol, 233

Packaging and thermal management, 8
Packet switching, 186
Parallel
   array processors, 123
   benchmarks, 65
   paradigm, 4
   processing architecture, 21
Partial benchmarks, 66
Parbegin, 259
Parend, 259
Paris, 214
Peak rate, 65
PERFECT, 68
Perfect shuffle, 194
Performance Evaluation, 64
Performance models, 263
Performance-to-cost ratio, 1
Pipeline
   asynchronous, 75
   arithmetic, 77

Pipeline (*contd.*)
  conditional branches, 102
  cycle time, 75
  dynamic, 84
  instruction, 77
  linear, 82
  model, 73
  multifunction, 84
  nonlinear, 82
  performance, 113
  performance evaluation, 113
  scalar, 85
  static, 84
  types, 77
  unifunction, 84
  vector, 85
Pipelined processing paradigm, 3
Pipelined vector processors, 123
Precedence constraints, 52
Precise interrupt, 107
Processing Nodes (PRN), 153
Processing paradigms, 52
Processor-time product, 202
Programmed I/O, 26
Programming
  DOALL, 259
  Fork and Join, 257
  models, 12
  parbegin/parend, 259
  send/receive, 260
  shared variables, 260
  SIMD, 204
  vector processors, 147
Programmer tools and environments, 10
Provisioning, 301
Pseudo processing vector, 154
Phthalocyanine, 325


RAS, 11
Radial Basis Function Networks (RBF), 307
Read-read forwarding, 99
Real world/Application benchmark, 66
Recurrent Neural Networks (RNN), 309
Reduced instruction set computers (RISC), 24, 56
Register-oriented vector processor architecture, 127
Register rotation, 31
Register set, 33
Regularity, 184
Reservation station, 36
R/C ratio, 223
Router, 212
Routing protocols, 185

Scalability, 202
Scalar
  expansion, 150
  pipeline, 85
  renaming, 149
Scheduling, 254
Selector channels, 27
Self-organizing maps, 309
Semantic gap, 17
Semaphores, 250
Send/receive, 260
Serial paradigm, 3
Serial-parallel-serial, 54
Shared memory architecture, 224
Shared memory MIMD, 60
Shared path, 183
Shared variables, 260
SIGNAL, 250
SIMD, 59
  organization, 167
SIMD/MIMD, 224
Single processor benchmark, 66
SISD, 58
Skewed storage, 175
SLALOM, 68
SPECmarks, 67
Speculation, 30
Speedup, 76
Stage(segment), 73
Staging register, 73
Stanford small programs, 68
Star, 188
Start-up time, 125
Static
  pipeline, 84
  scheduling, 254
  topology, 185
Store and forward, 186
Straight storage, 175
Strip mining, 139
Structural hazard, 86
Subprogram inlining, 152
Superpipelining, 25
Superscaling, 25
Sustained rate, 65
Switch controller, 183
Switch node, 439
Switching networks, 194
Sun Microsystem's Niagara microprocessor, 117
Supervisor/worker, 54
Supervisory node(SRN), 154
Synchronization mechanism, 372
Synapse, 306
Synchronous array processor, 167
Synthetic benchmarks, 67
System architectures, 10

# Index

System modeling and performance analysis, 11
System X, 274
SX-6, 158

Task, 50
Taxonomy
　Flynn, 55
　Feng, 185
TEST_AND_SET, 248
Thinking Machine Corporation
　Connection Machine-2 (CM-2), 207
　Connection Machine-5 (CM-5), 268
Tightly-coupled architecture, 63
Time-constrained scaling, 262
Token ring, 237
Tomosulo's Algorithm, 110
Translation Look aside Buffer(TLB), 32
Travelling salesman problem, 318
Trivial parallelism, 4, 54
Two-dimensional mesh, 188

UMA (Unform Memory Access Architecture), 225
Uniprocessor model, 15

Unix utility an application benchmarks, 66
Unlock, 248

Vector
　pipelines, 85
　processor, 123
　processor models, 124
　performance evaluation, 144
　programming, 147
VCSEL, 325
Virtual memory, 22
VLIW, 119
Von Neumann model, 16

WAIT, 250
Waves of computing, 2
Whetstone, 67
Wider word fetch, 21
Wormhole switching (cut-through routing), 186
Write-back, 230
Write-once, 230
Write-read forwarding, 99
Write-through, 230
Write-write forwarding, 99

Zero-address machines, 19